CAMBRIDGE LIBRARY COLLECTION

Books of enduring scholarly value

Botany and Horticulture

Until the nineteenth century, the investigation of natural phenomena, plants and animals was considered either the preserve of elite scholars or a pastime for the leisured upper classes. As increasing academic rigour and systematisation was brought to the study of 'natural history', its subdisciplines were adopted into university curricula, and learned societies (such as the Royal Horticultural Society, founded in 1804) were established to support research in these areas. A related development was strong enthusiasm for exotic garden plants, which resulted in plant collecting expeditions to every corner of the globe, sometimes with tragic consequences. This series includes accounts of some of those expeditions, detailed reference works on the flora of different regions, and practical advice for amateur and professional gardeners.

Ladies' Botany

The horticulturalist John Lindley (1799–1865) worked for Sir Joseph Banks, and was later instrumental in saving the Royal Horticultural Society from financial disaster. He was a prolific author of works for gardening practitioners but also for a non-specialist readership, and many of his books have been reissued in this series. The first volume of this two-volume work was published in 1834, and the second in 1837. At a time when botany was regarded as the only science suitable for study by women and girls, Lindley felt that there was a lack of books for 'those who would become acquainted with Botany as an amusement and a relaxation', and attempted to meet this need. The first volume, in the form of engaging letters to a lady, was originally intended to stand alone. Illustrated with detailed botanical drawings, it schools the student in botanical form and taxonomy as well as nomenclature.

Cambridge University Press has long been a pioneer in the reissuing of out-of-print titles from its own backlist, producing digital reprints of books that are still sought after by scholars and students but could not be reprinted economically using traditional technology. The Cambridge Library Collection extends this activity to a wider range of books which are still of importance to researchers and professionals, either for the source material they contain, or as landmarks in the history of their academic discipline.

Drawing from the world-renowned collections in the Cambridge University Library and other partner libraries, and guided by the advice of experts in each subject area, Cambridge University Press is using state-of-the-art scanning machines in its own Printing House to capture the content of each book selected for inclusion. The files are processed to give a consistently clear, crisp image, and the books finished to the high quality standard for which the Press is recognised around the world. The latest print-on-demand technology ensures that the books will remain available indefinitely, and that orders for single or multiple copies can quickly be supplied.

The Cambridge Library Collection brings back to life books of enduring scholarly value (including out-of-copyright works originally issued by other publishers) across a wide range of disciplines in the humanities and social sciences and in science and technology.

Ladies' Botany

Or, a Familiar Introduction to the Study
of the Natural System of Botany

VOLUME 1

JOHN LINDLEY

CAMBRIDGE
UNIVERSITY PRESS

CAMBRIDGE
UNIVERSITY PRESS

University Printing House, Cambridge, CB2 8BS, United Kingdom

Cambridge University Press is part of the University of Cambridge.
It furthers the University's mission by disseminating knowledge in the pursuit of
education, learning and research at the highest international levels of excellence.

www.cambridge.org
Information on this title: www.cambridge.org/9781108076579

© in this compilation Cambridge University Press 2015

This edition first published 1834
This digitally printed version 2015

ISBN 978-1-108-07657-9 Paperback

Selected botanical reference works available in the
CAMBRIDGE LIBRARY COLLECTION

al-Shirazi, Noureddeen Mohammed Abdullah (compiler), translated by
Francis Gladwin: *Ulfáz Udwiyeh, or the Materia Medica* (1793)
[ISBN 9781108056090]

Arber, Agnes: *Herbals: Their Origin and Evolution* (1938)
[ISBN 9781108016711]

Arber, Agnes: *Monocotyledons* (1925) [ISBN 9781108013208]

Arber, Agnes: *The Gramineae* (1934) [ISBN 9781108017312]

Arber, Agnes: *Water Plants* (1920) [ISBN 9781108017329]

Bower, F.O.: *The Ferns (Filicales)* (3 vols., 1923–8) [ISBN 9781108013192]

Candolle, Augustin Pyramus de, and Sprengel, Kurt: *Elements of the Philosophy
of Plants* (1821) [ISBN 9781108037464]

Cheeseman, Thomas Frederick: *Manual of the New Zealand Flora*
(2 vols., 1906) [ISBN 9781108037525]

Cockayne, Leonard: *The Vegetation of New Zealand* (1928)
[ISBN 9781108032384]

Cunningham, Robert O.: *Notes on the Natural History of the Strait of Magellan
and West Coast of Patagonia* (1871) [ISBN 9781108041850]

Gwynne-Vaughan, Helen: *Fungi* (1922) [ISBN 9781108013215]

Henslow, John Stevens: *A Catalogue of British Plants Arranged According to
the Natural System* (1829) [ISBN 9781108061728]

Henslow, John Stevens: *A Dictionary of Botanical Terms* (1856)
[ISBN 9781108001311]

Henslow, John Stevens: *Flora of Suffolk* (1860) [ISBN 9781108055673]

Henslow, John Stevens: *The Principles of Descriptive and Physiological Botany*
(1835) [ISBN 9781108001861]

Hogg, Robert: *The British Pomology* (1851) [ISBN 9781108039444]

Hooker, Joseph Dalton, and Thomson, Thomas: *Flora Indica* (1855)
[ISBN 9781108037495]

Hooker, Joseph Dalton: *Handbook of the New Zealand Flora* (2 vols., 1864–7) [ISBN 9781108030410]

Hooker, William Jackson: *Icones Plantarum* (10 vols., 1837–54) [ISBN 9781108039314]

Hooker, William Jackson: *Kew Gardens* (1858) [ISBN 9781108065450]

Jussieu, Adrien de, edited by J.H. Wilson: *The Elements of Botany* (1849) [ISBN 9781108037310]

Lindley, John: *Flora Medica* (1838) [ISBN 9781108038454]

Müller, Ferdinand von, edited by William Woolls: *Plants of New South Wales* (1885) [ISBN 9781108021050]

Oliver, Daniel: *First Book of Indian Botany* (1869) [ISBN 9781108055628]

Pearson, H.H.W., edited by A.C. Seward: *Gnetales* (1929) [ISBN 9781108013987]

Perring, Franklyn Hugh et al.: *A Flora of Cambridgeshire* (1964) [ISBN 9781108002400]

Sachs, Julius, edited and translated by Alfred Bennett, assisted by W.T. Thiselton Dyer: *A Text-Book of Botany* (1875) [ISBN 9781108038324]

Seward, A.C.: *Fossil Plants* (4 vols., 1898–1919) [ISBN 9781108015998]

Tansley, A.G.: *Types of British Vegetation* (1911) [ISBN 9781108045063]

Traill, Catherine Parr Strickland, illustrated by Agnes FitzGibbon Chamberlin: *Studies of Plant Life in Canada* (1885) [ISBN 9781108033756]

Tristram, Henry Baker: *The Fauna and Flora of Palestine* (1884) [ISBN 9781108042048]

Vogel, Theodore, edited by William Jackson Hooker: *Niger Flora* (1849) [ISBN 9781108030380]

West, G.S.: *Algae* (1916) [ISBN 9781108013222]

Woods, Joseph: *The Tourist's Flora* (1850) [ISBN 9781108062466]

For a complete list of titles in the Cambridge Library Collection please visit:
www.cambridge.org/features/CambridgeLibraryCollection/books.htm

LADIES' BOTANY:

OR

A FAMILIAR INTRODUCTION

𝔗𝔬 𝔱𝔥𝔢 𝔖𝔱𝔲𝔡𝔶

OF THE

NATURAL SYSTEM OF BOTANY.

BY

JOHN LINDLEY, Ph. D. F.R.S.

ETC. ETC. ETC.

PROFESSOR OF BOTANY IN THE UNIVERSITY OF LONDON.

Dich verwirret, Geliebte, die tausendfältige Mischung,
  Dieses Blumengewühls über dem Garten umher;
Viele Namen hörest du an, und imme verdränget
  Mit barbarischem Klang, einer den andern im Ohr.
Alle Gestalten sind ähnlich, und keine gleichet der andern;
  Und so deutet das Chor auf ein geheimes Gesetz,
Auf ein heiliges Räthsel. O! könnt ich dir, liebliche Freundinn,
  Ueberliefern sogleich glücklich das lösende Wort.—Göthe.

LONDON:
JAMES RIDGWAY AND SONS, PICCADILLY.

MDCCCXXXIV.

PREFACE.

This little book has been written in the hope that it may be found useful as an elementary introduction to the modern method of studying systematic Botany.

There are many works, of a similar description, to explain or illustrate the artificial system of Linnæus, the simplicity of which might have rendered such labours superfluous; but no one has, as yet, attempted to render the unscientific reader familiar with, what is called, the Natural System, to which the method of Linnæus has universally given way among Botanists. All seem curious to know something about this celebrated System, and many, no doubt, take infinite pains to understand it ; but it is to be feared, that a large part of those who make the attempt, are far from meeting with the success their industry deserves. On all hands they are told of its difficulties ; books, instead of removing those difficulties, only perplex the reader by multitudes of unknown words, and by allusions, which, however clear they

may be to the experienced Botanist, are anything
rather than illustrative in the eyes of a beginner,
who is often fairly lost in a labyrinth of resemblances,
differences, and exceptions. One would think modern Botany was like "the art unteachable, un-
taught," only to be understood by inspiration.

The cause of this lies, not in the science itself, so
much as in the books that are written concerning it.
Since the appearance of my *Introduction to the Natural
System of Botany* in 1830, several works of great
merit have been published on the same subject, both
in this country and abroad, so that the student is
abundantly supplied with guides ; and if his object
be to understand it, as an important branch of
Natural Science, they are sufficiently well adapted
to his purpose ; but for those who would become
acquainted with Botany as an amusement and a
relaxation, these works are far too difficult. Treating
the subject, as they do in great detail, and without
consideration for the unlearned reader, the language,
the arguments, and the illustrations employed in them
must be unintelligible to those who have no previous
acquaintance with Botany ; the characters of the
Natural Groupes or Orders, into which the Vegetable
Kingdom is divided, are not as a whole, susceptible
of such an analysis as a young student is capable of

following; and I can quite understand how the whole system may appear to be an unintelligible mass of confusion. It has, therefore, occurred to me that if, without sacrificing Science, the subject should be divested of the many real and of the still greater number of imaginary difficulties that frighten students, and if they could be taught to recognize the Natural tribes of plants, not by mere technical characters, but by those simple marks of which the practised Botanist exclusively makes use, a work in which such objects are attained might be found of some utility.

It is now admitted on all hands that the principles of the artificial system of Linnæus, which were so important and useful at the time when they were first propounded, are altogether unsuited to the present state of science ; and in the latest work that has been published in this country, upon that system, the learned and amiable author is forced to rest his defence of his still following it upon "the facility with which it enables any one, hitherto unpractised in Botany, to arrive at a knowledge of the genus and species of a plant." But if a system of Botany is to be nothing more than a contrivance to help those who will not master the elements of the science, to deter-mine the name of a plant ; and if it is really neces-

sary to have a mental rail-road on which such persons may be impelled without any exertion of their
own; then indeed the analytical tables of the French
are infinitely better contrivances than the Sexual
System: because if well executed they meet every
case and lead with certainty to positive results.

I have, however, been always at issue with the
Linnean school of Botany as to their system accom-
plishing even the little that it pretends to ; and if I
may be permitted to appeal to my own personal
experience of the difficulties of a beginner who is
unassisted by a tutor, (and few could have had fewer
difficulties to contend against than myself,) I should
say that it is totally opposed to such a conclusion. I
began with the Linnean system, which I was taught
to believe little less than an inspired production; I
had plenty of books compiled according to that
system to consult, and I was fairly driven to seek
refuge in the Natural System from the difficulties and
inconsistencies of that of Linnæus.

It seems to me that there is a confusion of ideas in
what is urged in favour of the Linnean system, and
that its theoretical simplicity is mistaken for practical facility of application. That the principles of
the Linnean system are clear, and simple, and easily
remembered is indisputable ; that student indeed

must be remarkably dull of apprehension, who could
not master them in a day. But is its application
equally easy? that is the point. When, for example, a
specimen of a Monopetalous plant has lost its corolla,
or when the stamens or pistils are absent, either accidentally, or constitutionally, as in Dioecious plants,
what Linnean Botanist can classify the subject of
inquiry? Or where a genus comprehends species
varying in the number of their stamens, as for instance, Polygonum, Salix, Stellaria, and hundreds of
others, who is to say which of the species is to determine the classification of the rest? or when this point
has been settled, how is the student to know what
passed in the mind of the Botanical Systematist?
The latter puts a genus into Octandria, because out
of ten species, one has constantly, and two occasionally, eight stamens, and he includes in the same
class and order, all the other species of the genus,
although they have five, six, or ten stamens. Suppose the student meets with one of the last, and
wishes to ascertain its name by the Linnean system,
he will look for it in Pentandria, or Hexandria, or
Decandria, where he will not find it. After wasting
his time, and exhausting his patience in a vain pursuit, he must abandon the search in utter hopelessness,
for there is no other character that he can make use

of as a check upon the first. At last some one will tell him that his plant is a Polygonum ; he turns to his book, wondering how he could have overlooked it ; and he finds Polygonum in Octandria. Should he inquire how this is, he will learn that his species belongs to Octandria, not because it is octandrous, *but because it is so very like other Polygonums that it cannot be separated from them, and they belong in most cases to Octandria.* This is the unavoidable answer ; and what does it really mean, except that it is not in consequence of its accordance with the system that the student's Polygonum is to be discovered, *but in consequence of its natural relation to other Polygonums ;* so that it is necessary to understand the Natural System, to make use of the Artificial System ! This is no exaggerated case, but one of common occurrence. It is undoubtedly true that in some books such inconvenience is guarded against by special contrivances ; but those contrivances form no part of the system.

Granting, however, for argument's sake, that these and other objections are overstated, and that the Linnean system does really facilitate the discovery of the class and order to which a plant belongs, let us next consider what advance towards the determination of the genus and species, or in other words the name of a plant, a student has really made, when the class and order are ascertained. If this argument were conducted, as in strictness it ought to be, with reference to the whole Vegetable Kingdom, it would be easy to shew that the student had in fact gained almost nothing that is of use to him ; but, in order to give the friends of the Linnean system every advantage in the discussion, let us see of what use it will be to him in regard to the few hundred plants that grow wild in England. For this purpose take the generic characters in Diandria Monogynia, as stated in Dr. Hooker's British Flora, a work in which the subject is treated with all the skill and perspicuity of which it is susceptible, and in which the Linnean system is seen to the greatest advantage. The characters are these :—

* Perianth double, inferior, monopetalous, regular.

1. LIGUSTRUM, *Linn.* Privet.—*Cor.* four-cleft. *Berry* two-celled, with the cells two-seeded.

** Perianth double, inferior, monopetalous, irregular. Seeds enclosed in a distinct pericarp (Angiospermous).

2. VERONICA, *Linn.* Speedwell.—*Cor.*four-cleft, rotate, lower segment narrower. *Caps.* two-celled.

3. PINGUICULA, *Linn.* Butterwort.—*Cal.* two-lipped, upper lip of three, lower of one bifid segment. *Cor.* ringent, spurred. *Germen* globose. *Stigma* large, of two unequal plates or lobes. *Capsule* one-celled, with the seeds attached to a central receptacle.

4. UTRICULARIA, *Linn.* Bladderwort.—*Cal.* two-leaved, equal. *Cor.*

personate, spurred. *Stigma* two-lipped. *Caps.* globose, of one cell. *Seeds* fixed to a central receptacle.

*** *Perianth double, inferior, monopetalous, irregular. Seeds four, apparently naked (closely covered by the pericarp, Gymnospermous).*

5. LYCOPUS, *Linn.* Gypseywort.—*Cal.* tubular, five-cleft. *Cor.* tubular, *limb* nearly equal, four-cleft, upper segment broader, and notched. *Stam.* distant, simple.

6. SALVIA, *Linn.* Sage or Clary.—*Cal.* two-lipped, tubular. *Cor.* labiate, the tube dilated upwards and compressed. *Filaments* with two divaricating branches, one only bearing a perfect single *cell* of an *anther*.

**** *Perianth double, superior.*

7. CIRCÆA, *Linn.* Enchanter's Nightshade.—*Cal.* two-leaved, but united into a short tube at the base. *Cor.* of two petals. *Caps.* two-celled; *cells* one-seeded.

***** *Perianth single, or none.*

8. FRAXINUS, *Linn.* Ash.—*Cal.* O, or four-cleft. *Cor.* O, or of four petals. *Caps.* two-celled, two seeded, compressed and foliaceous at the extremity. *Seeds* solitary, pendulous. (Some flowers without stamens).

9. LEMNA, *Linn.* Duckweed.—*Perianth* single, monophyllous, membranaceous, urceolate. *Fruit* utricular.

10. CLADIUM, *Schrad.* Twig-rush.—*Perianth* single, glumaceous. *Glumes* of one piece or valve, one-flowered, imbricating; outer ones sterile. *Fruit* a nut, with a loose external coat, destitute of bristles at the base.

This extract from the British Flora makes it evident that in determining to what genus a plant belongs, a great deal of inquiry beyond the discovery

that it has two stamens and one style, is indispensable.
The student must be acquainted with the meaning of
many technical terms, he must have his plant in
different states of growth, he must procure the fruit,
he must examine the interior of that part ; in short,
he must go through a long and careful examination,
which is entirely independent of the Sexual System.
In other and larger classes, such as Pentandria, Hexandria, Tetradynamia, Syngenesia, Gynandria, and
Monœcia, the length and difficulty of such an examination are vastly increased. Now I distinctly
assert that there is no difficulty in determining the
Natural Orders of plants greater than that of making
out the genera in the Linnean system. In fact it is
the very same thing, only with a different result : in
the one case it leads to the mere discovery of a name ;
in the other to the knowledge of a great number of
useful and interesting facts independent of the name.
This, which I hope will be evident from a perusal of
the following Letters, is so strongly felt by all Botanists of any experience, that they never think of
using the Artificial System themselves ; they only
recommend it to others.

There is, however, no mistake into which the
public is apt to fall much greater than the notion
that Botany is a science of easy acquirement. Like

all other branches of Natural History, it is far too complicated in its phenomena, and too diversified in form to be attainable as a science without long and attentive study; nevertheless a certain amount of it may be acquired without extraordinary application. The following pages will, it is hoped, explain sufficiently in what way this may best be done.

What I should recommend to those who take up this work with the intention of studying it is to begin with the beginning, to follow it in the same order in which it is written, and to procure for examination the very flowers that are named in it; they are in most cases within the reach of every one who lives in the country. The specimens should be carefully compared with the descriptions and plates; and when they are all remembered and understood, you will be a Botanist;—not a very learned one—but acquainted with many of the fundamental facts of the science, and able to prosecute the inquiry to any further point, and to study other and more scientific works with ease and advantage.

The course to be pursued by those who would push their inquiries beyond the information in the present work should be of this nature. They should read some Introduction to Botany, in which the modern views of structure and of vital action are well explained ; they should make themselves familiar with technical terms, which, although avoided in the following Letters, cannot be dispensed with in works of a more exact and scientific character ; they may at the same time perfect themselves in a knowledge of Natural Orders, by gathering the wild plants that are within their reach, comparing them with each other, and with the characters assigned to them in systematic works. Having thus provided themselves with a considerable amount of fundamental knowledge, they may apply themselves to the study of the Natural System in its great features. They will then, and not till then, be able to appreciate the various modifications of organization that connect one tribe of plants with another, and to understand the infinite wisdom and beautiful simplicity of design which is so visible in the vegetable world ; the just appreciation of which, through countless gradations of form, structure, and modes of existence, it should be the constant aim of the Botanist to demonstrate.

TABLE OF CONTENTS.

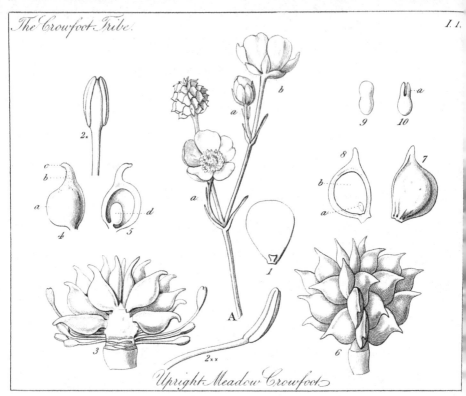

The Crowfoot Tribe. I. 1.

Upright Meadow Crowfoot.

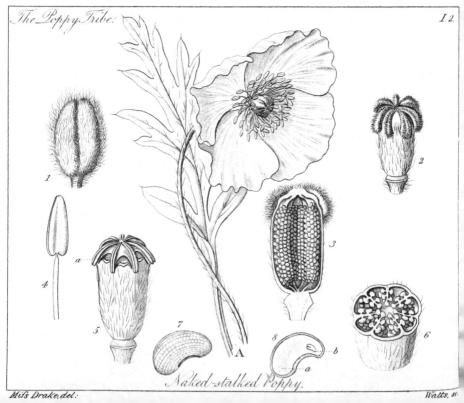

The Poppy Tribe. I. 2.

Naked-stalked Poppy.

LETTERS ON BOTANY,

&c. &c.

LETTER I.

INTRODUCTORY REMARKS—FUNDAMENTAL TERMS—
THE CROWFOOT TRIBE—THE POPPY TRIBE.

(Plate I.)

You ask me how your children are to gain a knowledge of Botany, and whether the difficulties which
are said to accompany the study of this branch of
science, cannot, by some little contrivance, be either
removed altogether, or very much diminished.—
Allow me, in answer to this question, to repeat a
fable which I remember to have read in some French
author.

A lady, observing some ants travelling across a
table, dropped a lump of sugar in the midst of them;
but, to her surprise, although ants are noted sugar-
eaters, they all retreated in terror from the spot, nor
could any of them afterwards find courage to return
to examine the object of their dread; on the contrary, they chose another track, and carefully avoided
that which would have proved a treasure had they
known its value. Struck by the occurrence, the lady

B

placed the same piece of sugar on a part of the table
near which the ants were in the habit of crossing,
and when she saw one of them approaching it, she
gently placed her finger in his way, so as to obstruct
his passage without alarming him ; the ant paused,
looked around him, and then took a new direction,
not exactly towards the sugar, but near it ; the
lady again opposed his passage gently, and at last,
by making him take a sort of zigzag direction,
tacking, as it were, at every few steps, the ant was
unconsciously brought to the sugar without being
frightened. Once there, he examined the glittering
rock attentively, touched it with his antennæ, broke
off a morsel, and hastened away with it to the ant-
hill ; thence he presently returned at the head of a
host of his comrades, by whom the rest of the sugar
was quickly carried off.

So it is with science and the young who have to
acquire a knowledge of it. Let them be once alarmed
at the aspect of their new pursuit, and it is almost
impossible to restore their confidence ; but there are
few who, if led to it insensibly, will not perse-
vere till they have made themselves masters of the
subject.

The most discouraging parts of Botany to a beginner, consist either in the numerous new and
strange names one has to learn the meaning of, or in
the minuteness of the parts by which plants are distinguished from each other, or in the great multitude of
species of which the vegetable kingdom consists; and
it must be confessed, that there is something seriously

alarming in the mass of preliminary knowledge which it would appear has to be acquired before any perceptible progress can be made.

But if we look at the subject a little more closely, we shall find that of the technical names employed, only a small number is really necessary in the beginning; that minute parts are little consulted in practice, however much they may be in theory; and that the arrangements of Botanists are so perfect, that no more inconvenience is experienced from the number of species, than in any other branch of Natural History.

There are certain terms, the exact meaning of which *must* be understood, and which cannot be dispensed with, if the science is to be studied to any good purpose; a sort of habit of observation has also to be acquired, without which the differences between one plant and another, can never be appreciated or remembered; but these things may be gained imperceptibly and without any extraordinary exertions either of industry or patience. We have only to begin with the beginning, and never to take one step till that which precedes it is secured; afterwards, the student may advance to what point he pleases. This appears to me to be the only secret in teaching Botany.

We must, however, be careful while we attempt to strip the study of its difficulties, that we do not also divest it of its interest, and imitate those who, by the ingenious substitution of words for ideas, have contrived to convert one of the most curious and interesting of all sciences into a meagre and aimless system of names.

Names are, no doubt, necessary for the purpose of dis-tinguishing one thing from another; but, as no one would call that man a carpenter, who only knew the names of all the instruments in a carpenter's chest of tools, so neither can that person be considered a Bo-tanist, whose knowledge is confined to the application of a few hundred Latin names to flowers and weeds. If the latter were the mark of a Botanist, then would every gardener be so accounted, which would be a great and grievous mistake.

" J'ai toujours cru," said Rousseau, whom I would at once have advised you to take as your guide, if his inimitable Letters were not both incomplete and obsolete,—" J'ai toujours cru qu'on pourrait être un très grand Botaniste, sans connaitre une seule plante par son nom,"—and to a great extent he was right. Only to apply their names to a few plants, is a poor insipid study, scarcely worth the following; but to know the hidden structure of such curious objects, to be acquainted with the singular manner in which the various actions of their lives are performed, and to learn by what certain signs their relationship, for they have their relations like ourselves, is indicated, is surely among the most rational and pleasing of pursuits.

It is only by acquiring a knowledge of the natural system of Botany, that you can systematically provide yourself with such information. If timid, or unin-formed, or interested persons have alarmed you with an account of the difficulties of this mode of study, let me advise you to disbelieve them, and to give your little people the opportunity of making the attempt. With such assistance as I shall be happy to give you, I cannot doubt of their succeeding.

You need not be told, that plants have generally five very distinct parts, viz. ROOT, STEM, LEAF, FLOWER, and FRUIT; the application of the three first of these terms, in their common acceptation, is already well known to you; the last is applied by Botanists, not only to such things as apples, pears, cherries, and the like, but also to any part which contains the seed; so that the grains of corn, the heads of the poppy, the nuts of the filbert, and even the little bodies which are commonly called caraways, are all different kinds of fruit. These terms may pass without further explanation for the present.

It is in the flower that the beauty of plants chiefly resides; it is there that we find all the curious apparatus by means of which they are perpetuated, and it is the spot where the greatest number of parts are found, the names of which are unusual, and require to be remembered. To illustrate these let us take a very common plant, to be found in every meadow, by the learned called Ranunculus, by the vulgar Buttercup, or *Crowfoot.*

On the outside of the flower of this plant, about the middle of its stalk, are one or two little leaves, which look like the other leaves, only they are a great deal smaller; indeed, they are so small as to resemble scales; these are the BRACTS (Pl. I. 1. *a. a.*).

Next them, and forming the external part of the flower itself, are five small greenish-yellow hairy

leaves (*A. b.*), which are rather concave, and fall
off shortly after the flower opens; leaves of this sort
form the CALYX, and are called SEPALS; it is supposed that they are intended to protect the more
tender parts of the flower, when the latter are very
young and delicate.

Next the sepals are placed five other leaves,
which are much larger, and of a bright shining yellow; they stand up and form a little cup, in the
bottom of which the other parts of the flower are curiously arranged; these five shining yellow leaves
form the COROLLA, and are called PETALS. It is they
that give the gay and glittering appearance to the
Crowfoot; which, when they drop off, is scarcely
to be distinguished from the grass it grows among.
Their business is, in part, to prepare the honey which
exudes from a little scale you will find on their inside, near their base (*fig.* 1.), and, which, if secreted
in sufficient quantity, is collected by bees for their
sweet food; and it is, in part, to protect from injury
the delicate organs which lie in their bosom. These
last are of two sorts; as you will soon learn.

In a ring from which both the sepals and petals
arise, you will find a number of little thread-like
yellow bodies, which are thicker at the top than at the
bottom; they spread equally round the centre, as if
they wished to avoid that part, and are a great deal
shorter than the petals; we call them STAMENS.
Their lower part, which looks like a thread, is called
the FILAMENT; their upper thickened end is named
the ANTHER. This last part is hollow, and will be

found, if you watch it, to discharge a small quantity
of yellow powder, called the POLLEN. The pollen
has a highly curious office to perform, as you will
learn presently.

Next to the stamens, and occupying the very
centre of the flower, are a number of little green
grains, which look almost like green scales; they
are collected in a heap, and are seated upon a small
elevated receptacle (*fig.* 3.); we call them CARPELS.
They are too small to be seen readily without a magnifying glass; but if they are examined in that way,
you will remark that each is roundish at the bottom,
and gradually contracted into a kind of short bent
horn at the top ; the rounded part (*fig.* 4. *a.*) is the
OVARY; the horn (*b.*) is the STYLE ; and the tip of the
style (*c.*), which is rather more shining and somewhat
wider than the style itself, is named the STIGMA ; so
that a carpel consists of ovary, style, and stigma.
At first sight, you may take the carpels to be solid,
and, if you already know something of Botany, you
may fancy them to be young seeds : but, in both
opinions, you would be mistaken. The ovary of each
carpel is hollow (*fig.* 5.) ; and contains a young seed
called an OVULE (*fig.* 5. *d.*), or little egg ; so that the
carpel, instead of being the seed, is the part that contains the seed.

Although the ovule is really the young seed, yet
it is not always certain that it will grow into a seed ;
whether or not this happens, depends upon the pollen,
of which we have already spoken, falling upon the
stigma. If the pollen does fall on the stigma, it

sucks up the moisture it finds there, swells, and finally each of the minute grains, of which it consists, discharges a jet of matter upon the stigma, which fertilizes the ovule, and then the latter grows and becomes a seed. But if the pollen does not fall upon the stigma, then the ovule withers away, and no seed is produced. Thus, you see every one of these parts of the flower is formed for some wise purpose. The sepals are to protect the petals; the petals to protect the stamens and carpels, and to form sugary food for their support; the stamens are to fertilize the ovules, and the carpels are to guard the young and tender seeds from injury; fertilization could not take place without the aid of the pollen; and the pollen could not produce its effect if it were not for the moisture and peculiar construction of the stigma. How admirable is the skill which is manifested in the construction of this little flower, and how striking a proof does it offer of the care with which the Creator has provided for the humblest of his works!

You have now seen all the parts of which flowers usually consist; the fruit is merely an alteration of the carpels, and the seed of the ovules. So perfect is the adaptation of the several parts to the end they have to perform, that it rarely happens that in the Crowfoot any of the ovules miss being fertilized. For this reason, the fruit of the Crowfoot is almost exactly the same when ripe as when young, except that its parts are larger, and it has become brown, dry and hard (*fig.* 6.). Separate, at this period, one of the carpels from the remainder, and place it under the microscope; it will be found to resemble a seed very much in appearance, and indeed is often called by that name ; but you have already seen, from an examination of the carpel, that the real seed is hidden in its inside ; formerly fruits of this sort were called naked seeds ; they are now called GRAINS ; as for instance a grain of caraway, a grain of wheat, and so on.

What remains to be seen of the structure of the Crowfoot is very minute, and requires some expertness in the use of a dissecting knife and microscope to be easily made out. It is not, perhaps, very important that you should understand it ; but, I may as well complete my account of the plant now that it is before us. I would recommend you to trust, at first, to the drawing that accompanies this ; and not to waste your time in cutting up the grains, until you have well understood what it is you have to look for.

The inside of the grain is filled up with the seed, now arrived at its perfect state ; the shell of the carpel has become hard and thick, and not only effectually protects the seed from harm (*fig.* 8. *a.*) but keeps it in the dark ; another wise provision, for without darkness the seed could not grow. The shell thus altered is called PERICARP.

If you cut the seed through, you will, for a long time, discover nothing but a solid mass of white flesh, in which all the portions seem to be alike ; but if you happen to have divided it accurately from top to bottom, cutting through both edges of the grain, as at *fig.* 8, you will then be able to discover, near the base of the seed, a very minute oval body *(fig. 8. a.)*, which may be taken out of the flesh with the point of a needle. This oval body is a young plant; it is the part which grows when the seed germinates, and is named the EMBRYO; the fleshy matter that surrounds it, called ALBUMEN, is only intended to nourish the young and delicate embryo, when it first swells and breaks through the shell. Small as is the embryo, so small as to be invisible to the naked eye, it also is organized in a regular manner. It is not merely an oval fleshy body, but it has two differently organized extremities, of which the one is divided into two lobes, called COTYLEDONS *(fig. 10.)*, or seed-leaves, and the other is undivided, and called RADICLE; the latter is the beginning of a root, as the former were the beginnings of leaves. Let the seed fall upon the earth, the embryo imbibes moisture, swells and shoots forth into a young Ranunculus, and thus the degrees of growth begin to be renewed as soon as they are completed.

Such is the structure of a perfect flower, and such the principal terms that you have to remember, in order to understand the language of Botanists; other terms there are, besides these, which are equally essential, but they are not used so frequently, and it is not worth detaining you about them now. I shall explain them whenever we meet with them.

Having thus minutely examined the flower of the Crowfoot, let us next observe the way in which its other parts are formed.

The ROOTS of this plant consist of a number of little taper divisions, the points of which are very tender and easily bruised ; from near the ends of these little divisions you have a number of delicate fibres, also with soft and tender points. It is by these points that the root obtains its watery food, from out of the soil; and if you break off the points the plant will languish until the wounds are healed, and new and perfect rootlets formed, or if it is unable to renew them it will die. This should always be thought of when you wish to transplant any thing ; for the same circumstance occurs in all other plants, and the success of removing shrubs, or flowers, or trees, depends very much upon the points of the roots being preserved. In the gay Asiatic Ranunculus, which is so often cultivated by gardeners, for the sake of the gaudy colours of its beautiful double flowers, at the time when the roots are taken out of the ground to be dried, all the fibres, up to the very points, have become so hard and tough as not to be easily injured ; yet it is possible even in this case to prevent the plants from producing healthy leaves, and well-formed flowers another season, if the roots are carelessly mutilated when taken up.

The LEAVES are dark-green, and very much divided into lobes, which are narrower in the leaves near the top of the stem than in those near the root. The business of these little appendages of the stem is not merely to render the face of nature pleasing to the eye by the charming verdure they produce ; nor, as in such plants as have eatable leaves, to supply men and ani-

mals with wholesome food. The plant itself could
not grow without them ; or, if it grew, could not
bear fruit, but would totally perish as soon as it was
born. The business of the leaves is to suck out
of the stem the watery food which the roots had
sucked out of the soil, and the stem out of the roots ;
and having filled themselves with it, to expose it to
light and to air, to evaporate the superfluous part,
and having thus in a manner digested it, to discharge
it again back into the stem in the form of the peculiar
matter which it may be the property of the plant to
produce ; such for instance as sugar in the Sugar
Cane, flour in the Potatoe, gum in the Cherry Tree, a
powerful medicinal substance in the Peruvian Bark
tree, and poison in the Ranunculus itself. In order
to enable the leaf to convey the watery food it sucks
out of the stem to all parts of its own surface, and to
return it back again, nature has furnished this little
organ with a most curious and complicated arrangement of drains or conduits, consisting of excessively
minute water-pipes glued together and branching in
every direction , these are what we call veins ; which
you see in the Ranunculus are all joined together in
the-stalk of the leaf, but separate as soon as they enter
the leaf itself, when they first divide into a few large
arms, then subdivide into a number of smaller ones,
and again divide over and over again, till at last they
form a net with finer and more delicate meshes than
the most exquisitely manufactured lace. If you could
examine the veins with a microscope you would find
them still more curious in their structure than I have

represented them to be ; for in the middle of every
little vein is a sort of windpipe which conveys vital
air to their extremest points.

Such is the way in which the veins of the leaf are
disposed in the Crowfoot. It is very essential that
you should pay attention to this netted branching
arrangement, because one of the great natural divisions of the Vegetable Kingdom is to be known by
that circumstance. One of the great natural divisions
of plants is called EXOGENOUS, because the stems
grow by addition to the outside of their woody centre ;
the same division is also called DICOTYLEDONOUS,
because the embryo has two seed-leaves or cotyledons.
You may generally recognize such plants by their
leaves having netted veins ; so that you see it is neither
necessary to watch the stem to see how it grows, nor
to examine the seed under a microscope in order to
count the cotyledons, if you would ascertain whether
a plant is Exogenous or not ; for the very obvious
arrangement of the veins in the leaves reveals the
secret structure of the stem.

If you have rightly understood what I have explained thus far, you will not only have mastered
several terms, but you will actually have made one
important step in the study you have taken up ;
you will have become acquainted with the essential
characters of a large and important natural order of
plants, called the CROWFOOT TRIBE (or Ranunculaceæ),
among which are some that are remarkable for the
virulent poison which often lurks beneath their beautiful exterior. The Crowfoot genus itself contains

some species, such as that called " the wicked" (Ranunculus sceleratus), and another called " the burning" (R. acris), which will blister the skin if applied externally to the human body, or produce dangerous symptoms if taken into the stomach; but these are mild and harmless if compared with such as I shall presently mention to you. But first of all let us see in what the most essential character of the Crowfoot tribe consists.

By far the greater part of the characters which we have seen that the Crowfoot possesses, will be also found in other and extremely different plants; but there are two characters that are what we call *essential:* that is to say, such as will distinguish it and the other plants belonging to the same natural order, from other natural orders, which resemble it. These essential characters are, there being a great many stamens which arise from beneath the carpels, (which is what Botanists term being *hypogynous*, Pl. I. 1. *fig.* 3.) ; and also several carpels which are not joined together. If you will pay attention to these two circumstances, you will always know a Ranunculaceous plant, that is to say, a plant belonging to the Crowfoot tribe ; and, although you may not know its name, you will know, what is of far more consequence, that it is, in all probability, a poisonous plant, and that its stem grows by addition to the outside of the woody centre, beneath the bark. A few instances of other genera, belonging to the same natural order, will put this in a clearer light.

There is a little annual, with leaves cut into divisions as fine as hairs, and with small but rich dark scarlet flowers, called Adonis, or *Pheasant's eye*. This plant agrees with the Crowfoot in its structure, in almost every point; but its petals have not a scale near their base, on which account Adonis is reckoned a plant of the same tribe, but is separated as a distinct genus.

Another and a charming little collection of pretty flowers is formed by the *Anemones*, with their purple, or white, or scarlet petals, which modestly hang their heads, as if unwilling to expose their beauty to every curious eye. These have the calyx and corolla mixed together, so that you cannot distinguish the one from the other; and when their flowers are gone, they bear little tufts of feathery tails, or oval woolly heads, in the place of the clusters of grains that you found in the Ranunculus. Such tails, or heads of wool, are collections of the grains of the Anemone, and contain the seeds; the tails themselves are nothing but the styles of the carpels, grown large and hard and hairy; they are thought to be intended by nature as wings, upon which the grains may be carried by the wind from place to place. If you look at the leaves, or the stamens, or the young carpels, or the ripe seeds of the Anemone, you will find all those parts constructed, in every essential respect, like the Crowfoot.—*Hepaticas*, which you have so often seen thriving, when neglected, in a cottage garden, when, perhaps, they perished under your own continual care, as if they were created specially for the pleasure of the poor, are nothing but Anemones, with

three bracts underneath the flower, and either six
or nine sepals and petals.

You have seen the *Globe-flowers* (Trollius), with
their yellow heads, in the borders among American
plants ; these have a great many sepals, which give
the beauty to their flowers; their petals are little
tubular bodies lying on the outside of the stamens ;
and each of their carpels contains several seeds; in
this they differ from the Crowfoots, but otherwise resemble them so much, that before the flowers appear,
you would take the Trollius for a Ranunculus.

Marsh Marigold, again (Caltha), which grows in
large green tufts in the meadows, and by the sides of
ditches, differs from the Globe-flower in having no
petals at all. *Christmas Roses* (Helleborus), and *Winter
Aconites* (Eranthis), are also of the Crowfoot tribe ;
each differing in one respect or other from those I
have mentioned, but possessing the essential characters already explained.

All the plants yet mentioned, are so like the
Ranunculus, that it is impossible to overlook their
resemblance, even if it were not pointed out. But
there are some other plants in which the resemblance is less striking, although it is equally great
when understood. Who has not heard of Larkspurs ?
Rocket Larkspurs, with their spikes of white, and
pink, and purple starry flowers ; of Bee Larkspurs,
which look as if the insect from which they take
their name were glued to their inside ; or Siberian
Larkspurs, with their branches of blue flowers, which
no gem nor mineral can emulate in brightness, or

deepness of colour? These are of the Crowfoot tribe, but less closely allied to Ranunculus itself, than what I have already mentioned. We are so accustomed to combine in our minds the idea of Buttercups and yellowness; that we are apt to overlook the resemblance that really exists between the Ranunculus and the Larkspur, because of the want of yellow in the latter. But setting aside this, which is of no Botanical importance whatever, let us look at the calyx of the Larkspur (Delphinium). It is composed of five leaves, or sepals, the uppermost of which has a horn arising from out of its back; so is that of Ranunculus, excepting the horn. It has four petals, of which two have long tails, hidden within the horn of the petal; it is they which look like the bee's body: Ranunculus has nothing of this; but five common petals instead. The Larkspur has a great many stamens arising from below the carpels; this is the *first essential* character of Ranunculus; it has also several carpels (two or three), which are not grown together : and this is the *second essential* character of Ranunculus; so that this plant has, in reality, no essential character by which it can be distinguished from the Crowfoot tribe. As for its four strange petals, they are of no importance; for you will remark, that in Trollius, the petals are little hollow things, and that the Marsh Marigold has none whatever; so that not only the form of the petals is of no consequence in the Crowfoot tribe, but it does not even signify whether they are present at all or not.

When you have once satisfied yourself that the

c

Larkspur belongs to Ranunculaceæ, you will have no
difficulty in perceiving that the true *Aconite* (Aconi-
tum), is also of the same natural order; for the resemblance of these two is too striking to be mistaken.
The true Aconite, which yields to no plant in the virulent poison of its roots, has all the structure of the
Larkspur, except that the upper leaf of its calyx has
not any horn, but is very large, and resembles a
sort of helmet, overshadowing all the other parts of
the flower. In consequence of there being no horn
to the upper leaf of the calyx, the two uppermost
petals, which have horns, are forced to hide them beneath the helmet, instead of inserting them into it.
With these differences the Larkspur and the true
Aconite are formed nearly alike.

Pæonies are the last plants I need mention, as belonging to the Crowfoot tribe. They have a calyx
which resembles green leaves, and which nèver drops
off; in other Ranunculaceæ, the calyx often drops off,
even before the petals ; in this respect, Pæonies are
unlike the rest of the Crowfoot tribe; but they have
the two essential marks of distinction in the carpels
and the stamens.

Let me recommend you to procure, if possible, all
the plants that have thus been enumerated, and to
compare them with one another till you fully understand their resemblance, which you may very readily
do ; and then you will find, that to know the structure of the common Crowfoot, is, indeed, as I said in
the beginning, to know the properties and general
character of a large natural order.

The next plants I would advise you to study, are the *Poppies;* that singular genus, which, in the form of a few red flowered kinds, is the plague of the farmer, who calls them Redweed, and, in the form of another species, is rendered, by the folly and vice of man, the scourge of half the world, in the shape of opium. The Poppies form a genus of plants which represents the characters of a small natural order, very nearly related to the Crowfoot tribe, from which it differs in its properties, being of a stupifying, instead of a burning and blistering nature, as well as in its botanical characters. Like the Crowfoot, the Poppy has leaves with netted veins, and also a great many stamens, arising from under the carpels ; but, unlike the Crowfoot, the carpels are not several, and distinct from each other, but have all grown together into a single ovary (Pl. I. 2. *fig.* 2.) ; the styles are wanting; and the stigmas are elevated hairy lines, which spread equally from the top of the ovary, forming a sort of starlike crown. If you open the ovary, you will find that it consists of but one cell, or cavity ; and that several little plates, which project from the sides of it into the cavity *(fig.* 6.), are covered with very numerous and very small ovules, or young seeds. In course of time, the ovary changes to a hollow box, with a hard brittle shell, called a CAPSULE, which is the fruit: and when it has become of a pale brownish colour, it, and the seeds it contains, are ripe ; in this state it is the poppy-head you see in the windows of druggists' shops. So hard is the shell of the capsule, and so small are the seeds, that the latter would never be able to get out unless nature had contrived some certain manner of opening the box. Lid it has none, for the hardened stigmas bind down the top and prevent its opening : but in order to remove this impediment, a number of little valves open underneath the edges of the stigma (*fig.* 5. *a.*), and through these the seeds fall out.

You see then, that the Poppy differs from the Ranunculus, in having the carpels united into an undivided ovary, instead of being all separate ; in the same way all the Poppy tribe differ from all the Crowfoot tribe ; but this is not all. If you break the stem or leaf of a poppy, a milky fluid runs out ; in which the stupifying principle of the plant is contained ; no milk is found in the Crowfoot tribe, the juice of which is always watery and transparent. This is another and most important mark of distinction.

The most essential differences between the Crowfoot tribe, and the Poppy tribe, therefore, stand thus :—

Crowfoot Tribe, or Ranunculaceæ.

Stamens very numerous. Carpels distinct. Juice watery.

Poppy Tribe, or Papaveraceæ.

Stamens very numerous. Carpels united into one central ovary, with a single cavity. Juice milky.

Besides these marks, by which you may know the two natural orders, there are others. The Poppy has two sepals, and twice two petals ; this is never

found in the Crowfoot tribe; but as some of the
Poppy tribe have three sepals, and twice three petals,
the number of those parts is not sufficiently constant
to form an essential mark.

Having thus examined the flowers, and ascertained
what the differences and resemblances are between
the one tribe and the other, I would next advise
you to compare the leaves of the wild Poppies with
those of the Crowfoots, and you cannot fail to be
struck with their great resemblance; and thus will
you have become acquainted with two striking
natural orders, neither of which will you ever be
likely, I should hope, to forget.

For the sake, however, of impressing the characters
of the Poppy tribe more strongly upon your mind, I
should advise you to take a few more examples;
especially as the fruit, by which the Poppy tribe is
partly known, frequently seems, until it has been
carefully explained, to be very unlike the only
instance we have, as yet, examined.

On the sea-shore near Brighton, and indeed on
most parts of the English coast, there grows a blue-
leaved plant, which looks as if the salt of the sea
spray had encrusted itself upon its skin; but which
is sure to be remarked when in flower, because of its
large bright yellow flowers, and when in fruit by the
long stiff horns it bears in the place of a poppy head;
for this last reason it is called *the Horned Poppy*,
(Glaucium). This plant also belongs to the Poppy
tribe. It differs from the Poppy itself, in its fruit
being very long and slender, instead of short and

round ; but like that plant it contains only one
cavity.　The reason why the poppy head is so thick,
is, that it is formed out of as many carpels grown
together, as there are stigmas ; in the naked-stalked
Poppy there are seven ; in the Opium Poppy
(Papaver somniferum) there are a great many ; in
others more and in others less : but in the Horned
Poppy the fruit is formed of only two carpels grown
together, which will at once explain why the fruit is
so narrow.

Celandine (Chelidonium majus), a pale-green cut-
leaved plant, with little yellow flowers, found in groves
and in shady lanes, or in churchyards, by no means
uncommonly, is another plant belonging to the
Poppy tribe.　Its milk is of an orange colour, and has
a nauseous taste ; its fruit is a long pod constructed
like that of the Horned Poppy, but dropping when it
is ripe into two pieces called VALVES.

The Argemone or *Prickly Poppy*, and the Eschscholtzia, so very remarkable for its finely cut
bluish leaves, and bright yellow flowers, are also of
this natural order.　In the Eschscholtzia a circumstance happens which you should not omit to note,
because it seems to explain several things in other
plants, which appear at first sight very puzzling.
The flower of this plant before it expands is enclosed
in a taper-pointed green sheath, shaped like a hutkin,
which is pushed off by degrees as the petals unfold,
and at last drops to the ground.　What is this singular
part, which is so unlike any thing in the plants
hitherto examined ? it is the calyx, which, like that

of the Poppy, is formed of two sepals, but the sepals grow so firmly together by their edges, where they touch each other, that, when the time for them to fall off arrives, they are unable to separate; but as it is absolutely indispensable to the plant that the calyx should in some way or other be got rid of, in order to enable the flower to expand, nature has provided the calyx with the means of separating from the stalk by its base; and thus it is pushed off in the manner I have mentioned. This, which is the first instance you have yet seen, of two parts that stand next to each other having grown together, is an exceedingly common occurrence in plants, and is one of the means by which the real nature of flowers is frequently so masqued that one can hardly discover how it is they are formed. I would advise you to recollect this occurrence, and in good time you will see how it explains other things.

This letter will have found occupation enough for your little girls in beginning their Botany, without my carrying the subject further for the present.

EXPLANATION OF PLATE I.

1. THE CROWFOOT TRIBE.—A. A twig of the upper part of the stem of the *Upright Meadow Crowfoot; a a* bracts, *b* calyx.—1. A petal seen in the inside with the scale at the base.—2* An anther with a part of the filament, seen in front; 2** the same more magnified, viewed sidewise.—3. The centre of a flower cut through, the calyx and corolla being removed; the stamens are seen spreading outwards, with

their filaments originating from underneath the carpels; the latter occupy the centre, and are shewn to arise from a short conical central part, which is their receptacle.—4. One of the carpels; *a* the ovarium; *b* the style; *c* the stigma.—5. The same carpel cut open so as to shew the young seed, or ovule, *d*.—6. A cluster of ripe carpels, or grains.—7. One of the grains separate; compare this with fig. 4, on the opposite side of the plate.—8. The same grain cut in half, shewing *a* the young plant or embryo, and *b* albumen or nutritive matter stored up for feeding the young plant when it begins to grow.—9. Is an embryo extracted from the albumen and seen from the back.—10. The same seen from the side, so as to shew the two minute seed-leaves or cotyledons, *a*. (N.B. The principal part of these figures is more or less magnified.)

2. THE POPPY TRIBE.—A. A flower and leaf of the *naked stemmed Poppy*, the flower of which is turned so as to exhibit the 4 petals, the stamens, and the ovary; the natural size.—1. Is a flower-bud, with the two sepals by which it is covered.—2. An ovary, with its diverging starlike stigmas.—3. The same part cut through from top to bottom, with the ovules or young seeds exposed.—4. A stamen, with the filament and anther.—5. A capsule, or head, when ripe; the little valves by which the seeds fall out are seen at *a*.—6. The capsule cut across, so as to shew the plates which project from its shell into the cavity, and the multitude of seeds that grow upon the plates.—7. A seed.—8. The same cut through; *a* the albumen; *b* the young plant, or embryo.

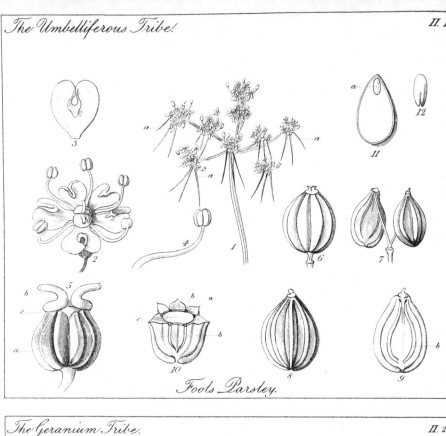

The Umbelliferous Tribe.

II. 1.

Fools Parsley.

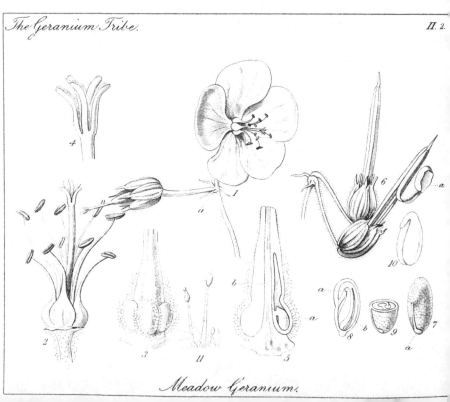

The Geranium Tribe.

II. 2.

Meadow Geranium.

LETTER II.

(Plate II.)

You ask me whether the Strawberry is not a plant
of the Crowfoot tribe; and if so, how it happens that
it is so wholesome a fruit. It is true, that the Strawberry plant has, in many respects, a resemblance to
the Crowfoot, especially in its numerous stamens
and carpels; and I, therefore, am not surprised at
your thinking that they must both belong to the
same natural order. But when you are more accustomed to the making accurate observations, you will
cease to be deceived by such resemblances. On a
future occasion I shall introduce the Strawberry and
its relations to your notice (see Letter VIII.); for the
present I shall content myself with calling your
attention to a difference by which they may be
readily distinguished. The Crowfoot is, as you have
seen, a plant with the stamens *arising from underneath
the carpels,* so that when the calyx drops, or is taken
off, the stamens still remain surrounding the carpels, and are, as Botanists say, hypogynous (see
p. 14.). But in the Strawberry flower, you cannot
tear off the calyx, without bringing the stamens

away along with it; the stamens therefore *arise out of the calyx*, which is a very different affair from what occurs in the Crowfoot, and in this way they, and the natural orders to which they respectively belong, may be certainly known from each other. When stamens have such an origin as in the Strawberry, they are technically called *perigynous*, a term which is so commonly used, that I could wish you to remember it.

Let me now shew you a natural order of plants, which, although less beautiful than the Poppy tribe and the Crowfoot tribe, is not less interesting or important. Enclosed with this letter, are fresh gathered specimens of a little plant with dark-green leaves, cut into many fine divisions, like parsley-leaves; their smell is strong and unpleasant. The flower stem of this weed is about a foot high, and it bears small greenish-white blossoms, arranged at the ends of the branches in a very peculiar manner (Pl. II. 1. *fig.* 1.). From its resemblance to Parsley, and from its noxious qualities, for it is very poisonous, the vulgar call it *Fool's Parsley* (Æthusa Cynapium). You will find it growing wild in almost any piece of waste garden ground, where it has sometimes been mistaken for Parsley that has sprung up spontaneously, and has proved fatal to unfortunate children who have eaten its leaves.

As it is rather a dangerous neighbour, you will not be sorry that I have taken an early opportunity of shewing you how to know the traitor, notwithstanding his resemblance to one of the most harmless of plants. Look at the way in which the flowers are arranged. You will remark that the flower-stem divides at the top into a number of short slender branches, which all proceed from a common point, just as the rays of a parasol all proceed from the ring which slides up the stick ; this is called an UMBEL. Each of these rays is terminated by a cluster of flowers, the stalks of which also proceed from a common point, and form again an umbel. So that the Fool's Parsley not only has its flowers arranged in umbels, but in what, in consequence of their divided nature, are called *compound* umbels ; the clusters of flower-stalks by which the branches of the first set of rays are terminated, are called *simple* umbels, by way of distinction. It is to this circumstance, of the peculiar arrangement of the flowers, that the natural order to which the Fool's Parsley belongs is called *Umbelliferous*, or Umbel-bearers ; for all the other plants, belonging to the order, have their flowers arranged in a similar manner ; and thus you have one exceedingly simple and manifest character by which the order is known.

Let us next look at a separate flower ; the parts are very small, but your little people's eyes are, I dare say, good enough for them to see all that I think it necessary to point out, without the use of a microscope. In the first place, there is scarcely any sign of a calyx ; all that you can find is a little narrow border, from within which the petals arise. The petals are five in number (*fig.* 2.), of a greenish-white colour, and each has its point doubled inwards (*fig.* 3.), in such a way as to look like the spots in a suit of hearts in playing cards; there are only five stamens, and they arise from between the petals. The centre of the little flower is occupied by a whitish fleshy body, which is divided into two lobes, from the top of each of which springs a style (*fig.* 5.). Where now is the ovary? Is it the whitish fleshy body from out of which the styles arise? if so, young seeds must be found in the inside of that part; but if you cut it, you will find neither cavity nor young seed.

The ovary is in this plant so concealed, that a beginner would not be likely to find it without assistance. Look at the top of the flower-stalk, on the outside of the petals; you will find a thickish deep-green furrowed part (*fig.* 5. *a.*), from the top of which the petals and stamens spring; and if you cut that part across you will discern two little cavities, in each of which hangs a young seed. This then is the ovary, which in Umbelliferous plants seems placed below the calyx and the corolla, on which account it is called *inferior*, just as in the Crowfoot and Poppy tribes, in which it stands above the calyx and corolla, it is termed *superior*.

These terms, like many others used in Botanical books, were invented long since, when Botany was in a rude state, and they convey an incorrect notion of the true nature of the parts to which they are applied. I shall, therefore, digress a little, to explain to you the real nature of the difference between a superior and an inferior ovary. In plants, such as the Crowfoot, which have a superior ovary, the sepals, the petals, the stamens, and the pistils, grow *separate* from each other; but in umbelliferous plants, whose ovary is inferior, the principal part of the calyx, the stalks of the petals, and the lower half of the filaments, grow to the sides of the ovary so firmly that they cannot be separated; and hence the ovary looks as if it grew beneath the other parts; while, in reality, the parts of the flower, both of the Crowfoot and Fool's Parsley spring from the top of the flower-stalk, beneath the pistil; but in the former, they are all separate; while in the latter, they grow all together.

After the petals and stamens have fallen off, the ovary gradually increases in size ; the furrows on its surface become deeper ; it hardens, and changes to a dull brown, and, at last, a fruit (*fig.* 6.) is formed, which, in time, separates into two halves (*fig.* 7. *a. b.),* or grains ; which are what are vulgarly and inaccurately named seeds.

These are the principal peculiarities in the flowers of umbelliferous plants; and they will always serve to know them by. You may, if you wish it, render their characters more simple and more easy to remember, by taking the essential distinction of umbelliferous plants to consist in *flowers growing in umbels, and inferior fruit, which when ripe separates, or may be separated, into two grains.* The appearance of this order is, however, so peculiar, that I have no fear of its being recognised with certainty after you have seen a few more instances of it.

Let us now return to the means of distinguishing

Fool's Parsley from common Parsley. Observe, once
more, the simple umbels of the former species; at the
bottom of the flower-stalks there are a few narrow
taper-pointed green leaves, which you will, probably, recognize by their situations to be bracts
(see page 5. and *fig*. 1. *a. a. a.*). When bracts surround a number of flowers in a ring, just as sepals
surround petals, and petals surround stamens, they
form what is called an INVOLUCRE ; it is, therefore,
by this name that I must speak of the bracts of umbelliferous plants. The involucres, then, of Fool's
Parsley, consist each of three leaves, which all turn
one way, spreading towards the outside of the umbel;
by this easy character it may be certainly known
from common Parsley, and from all the rest of
our wild umbelliferous plants, the involucres of which
are quite different.

It would be a good thing if they could all be distinguished from each other as easily ; but, unfortunately, umbelliferous plants are often so much alike,
that nothing but a very minute attention to the formation of the fruit, will enable you to find out their
names and qualities with certainty. I shall not pretend to shew you much of the manner of doing this;
for, if you would learn to distinguish them, you
must peruse books on systematic Botany, in which
umbelliferous plants are described. But, as it will
be useful for you to know the meaning of some
words that are, of necessity, employed in speaking
of their fruit, I may as well explain what those are.

The back of each half of the fruit of Fool's Parsley,

has four deep *furrows* (valleculæ) which cause five
elevated *ridges* to appear *(fig.* 10. *a.)* ; the nature of
these furrows and ridges is very much attended to.
The faces of the halves of the fruit, where they
touched each other before they were separated, are
also noticed ; they are called the *commissure.* In the
skin of the seed, underneath the rind of the fruit,
you will often see very minute slender brown lines
(fig. 10. and 9. *b.),* which, if highly magnified, are
found to be bags filled with oil ; they are called
stripes, or *vittæ,* and it is in them that the substance
which gives so pleasant a flavour to caraways and
coriander grains, is stored up. You may, perhaps,
have a difficulty in finding the stripes ; if so, cut a
ripe grain across *(fig.* 10.), and you will see the
ends of the stripes *(fig.* 10. *b.)* looking like little
mouths, out of which a dark oily matter slightly
oozes ; or, if you are still unsuccessful in your search,
then make a thin slice in the same direction, place
it in water under the microscope, and throw light
upon it from beneath, by means of the mirror, and
the ends of the stripes will appear as so many holes.

There are few tribes of plants more familiar to us
than the umbelliferous, because of the many useful
species that it contains. The carrot, the parsnep,
celery, eryngo, angelica, lovage, caraway, coriander, dill, anise, hemlock, fennel, and samphire,
are all well known kinds ; besides which there is a
host of others with which the Botanist is acquainted.
I should advise you to desire the gardener to find
you specimens of all those I have mentioned, and,

on learning the distinctions of them, to compare
them with each other, and with the descriptions of
Botanists, before you attempt to attend to the wild
species; in that way you will familiarize yourself
a little with the manner in which the characters are
drawn up ; a very necessary qualification for every
one who would study umbelliferous plants like a
Botanist.

The greater part of the species has white or whitish
flowers : a good many, as the fennel for example,
have yellow flowers, and a very few blue ones.   To
the latter class belong most of the species of Eryngo,
and the beautiful Didiscus cœruleus (*Bot. Register,
fig.* 1225.), which, although a native of New Holland,
forms so charming a hardy ornament of the flower
garden in the summer.

It is not a common circumstance for a tribe so
very similar in the structure of the species as umbelliferous plants, to contain both poisonous and
wholesome kinds; but here we have the Deadly
Hemlock and Dropwort, associated with Parsley,
Carrots and Fennel; and what seems still more remarkable, a species, Celery, which is unwholesome
in its wild state, become harmless when cultivated.
Common Celery is a native of the meadows of many
parts of England, where it forms a rank weedy
strong-smelling herb, which is unfit for human food ;
how different it is in gardens every body knows.   It
is thought that its ceasing to be noxious when cultivated is owing to the greater part of its stems and
leaves being blanched.   No doubt you must be

curious to know why blanching a plant should destroy its unwholesomeness, and therefore we will again digress from our principal subject for the purpose of explaining this curious fact.

In my last letter, I told you that the business of leaves is to expose to light and air the sap they suck out of the stem. The consequence of light and air acting upon the surface of leaves, is the forming in their substance, which is originally of the same yellowish-white that you see in seeds, a green colour, which is more or less deep in proportion to the degree in which the light is powerful; thus a plant which stands exposed to the sun all day long, has its leaves of a darker green than another which grows among other trees, or near a building which throws it into the shade a part of the day: and the latter again is darker green than a plant which grows at the north side of a high wall, or in an enclosed court which the sun's rays never enter. In like manner, if you cause a plant or any part of a plant to grow in total darkness, it will be entirely destitute of greenness; or in other words, the substance of the plant will remain of its original yellowish-white, because no green matter can be formed but by the action of light; and if a part already green is kept for a long time in darkness, it will become yellowish-white, in consequence of all its green being destroyed by the peculiar action of the atmosphere upon plants in darkness.—This is the explanation of blanching. But mere loss of colour is not the only consequence of plants being kept in the dark;

you have already been told that poisons, when it is the nature of plants to yield poisons, are also formed in leaves by the action of light; the absence of this wonderful agent will therefore prevent the formation of poison, as well as the formation of green colour; and hence blanching renders poisonous plants harmless. Thus, in the Celery, but a small portion only of the leaves is exposed to light; the whole of the stem and of the lower part of the leaves is buried in the earth; the small quantity of noxious matter that might be formed by the few leaves which are allowed to bask in the sun, has to pass down the buried stalks of the leaves before it can reach the stem, where it would be laid up; but you know the leaf-stalk of the Celery is very long, and any thing which has to filter from the upper part of such a leaf to its bottom, has to take a long journey, in the course of which it is constantly under the destroying influence of darkness; so that before it can reach the stem, it will all have perished. A similar effect is produced by the Italians upon Fennel, which, although not a poisonous plant, has too powerful a taste to be a pleasant food, except as an ingredient for flavouring sauces. The Italians, in their warm climate, cause Fennel to grow rapidly in darkness, and thus obtain it in a state very like Celery in appearance; the darkness destroys the principal part of the flavour, no more of the Fennel taste being left than is sufficient to give the blanched stems a pleasant aromatic quality.

There are no plants which you are likely to mis-

take for umbelliferous, if you will only attend to
the exact nature of the characters I have already
explained. Nevertheless, as it is exceedingly difficult for beginners to comprehend the necessity of
exactness in natural history, and as you have already
been puzzled about the Strawberry and the Crowfoot,
I may as well caution you against an error which
you may fall into ; and I cannot do better than point
it out in the words of Rousseau himself. " If you
should happen, after reading my letter, to walk out
and find an Elder-bush in flower, I am almost sure
that at first sight you would exclaim ; here we have
an umbelliferous plant. You would find a large
umbel, a small umbel, little white blossoms, an inferior ovary, and five stamens ; yes, it must be an
umbelliferous plant. But let us look again ; suppose
I take a flower. In the first place, instead of five
petals, I find a corolla, with five divisions, it is true,
but nevertheless with all five joined into one piece ;
now flowers of umbelliferous plants are not so constructed. Here indeed are five stamens, but I see no
styles : I see three stigmas more often than two ;
and three grains more often than two ; but umbelliferous plants have never either more or less than
two stigmas, nor more nor less than two grains for
each flower. Frequently the fruit of the Elder is a
juicy berry, while that of umbelliferous plants is dry
and hard. The Elder, therefore, is not an umbelliferous plant. If you now go back a little, and look
more attentively at the way in which the flowers
are disposed, you will also find their arrangement

only in appearance like that of umbelliferous plants.
The first rays, instead of setting off exactly from the
same centre, arise some a little higher, and some a
little lower ; the little rays originate with still less
regularity ; there is nothing like the invariable order
you find in umbelliferous plants. In fact, the arrangement of the flowers of the Elder is in a *cyme*,
ana not in an umbel. See how mistakes will sometimes lead us to the discovery of truth."

You must not suppose from so much having been
said about the umbel, that that kind of arrangement
of flowers, always indicates an umbelliferous plant ;
on the contrary it is only when the umbels bear inferior fruits separating into two grains that they really
belong to plants of this natural order. There are
many other plants which bear umbels, with a different
structure of the flower, as for instance the *Geranium
tribe* (Plate II. 2.).   Many species of that natural
order have simple umbels, but the structure of their
flowers is exceedingly different. Suppose I fill up my
letter with some account of them ; if you are weary of
study for the moment, you can leave off at the point
at which we are now arrived, and resume the subject
on another day.

Do you know what a *Geranium* is ? if you ask the
gardener for one, he will bring you a neat-looking
shrubby green-house plant, with fragrant leaves and
upright umbels of beautiful red flowers ; this however
is not a Geranium, although it is nearly related to one,
and belongs to the Geranium tribe ; it is a Pelargonium.   The real Geraniums are little herbs which

grow by the way side, or in waste places, or in the meadows, and some of which are often cultivated in the borders of the flower-garden or shrubbery, for the sake of their gay red or white or purple blossoms. One of them, the Meadow Geranium, is so very common that you can scarcely fail to procure it; if you should fail, then almost any other kind will do as well for the purpose of enabling you to follow me.

This plant has roundish leaves, divided into several deep lobes, with the veins branching in the manner of Exogenous plants (p. 13.), a circumstance which also occurred in the umbelliferous order, although I forgot to mention it. The leaves are placed upon long hairy stalks which are singularly swollen at the base, where there grows a pair of pale green thin scales called STIPULES; these parts you have not before seen; they are frequently not met with in any species of a whole natural order, but when they do occur they usually accompany the leaves of every plant in the order; their use is unknown. The flowers of this species can hardly be said to be in an umbel, for only two of them grow together, but if more were to appear, as is the case in other species, they would all diverge from one common centre; and this you know would make an umbel. Their calyx consists of five ribbed sepals which spread when the flower is open, and when the petals have fallen off contract round the young and tender ovaries, to which they form an efficient protection. The petals are five, of a purplish-blue colour; they are very round at their ends, and spread in such a manner as to form a figure something resembling a saucer; their veins are unusually prominent, and give the petals a streaked or pencilled appearance.

We did not see any veins in the petals of the Crowfoots, or the Poppies, or the Umbelliferous plants, and yet we might have found them if we had paid attention, for veins are as regularly found in petals as in leaves, and, what is very curious, they have the same structure, except that they are usually composed of air-vessels only. The tough and flexible tubes, or water-pipes, which you find surrounding the air-vessels in the veins of leaves (p. 12.), are in those organs indispensable for the protection of the air-vessels, and for giving strength to the leaf, during the many months which it has to exist ; but all this wonderful provision against injury, would be thrown away in the petals, which never live beyond a few days, sometimes only a few hours; and would be prejudicial to that delicate and transparent appearance in them which we so much admire. Nature, therefore, who creates nothing in vain, has generally formed petals with veins composed of air-vessels only; and hence the extreme delicacy of the fragile corolla. The petals of the Geranium are so well adapted to shew you this arrangement, and you must be so curious to witness the way in which the secret workings of vegetation take place, that I am sure you will thank me for teaching you how you can best view the structure. For this purpose place the petal of a Geranium upon a piece of perfectly smooth and flat glass, such as is usually furnished for the transparent stage of a microscope ; wet it with water ; and then lay over it another flat piece of glass. Press the two glasses firmly together, and by degress you will squeeze all the air out of the petal, and it will become transparent. You may then, with a pretty good magnifying power, observe all the air-vessels of the veins distinctly, looking like fine threads of silver wire twisted up like a spiral spring. It is on account of this appearance that the air-vessels are called technically SPIRAL VESSELS.

The stamens (*fig.* 2.) are ten, arising from beneath the pistil : that is, are hypogynous ; they are placed in two rows, each of which consists of five stamens. The lower part of the filament is broad, and rather convex ; it curves a little towards the pistil, and then tapers off into the part which bears the anther.

The pistil (*fig.* 3.) has a very singular appearance. At the base it has five roundish projections, covered all over with clammy hairs ; from the top of these projections a sort of column arises, which at the point is five-lobed (*fig.* 4.); the projections are the ovaries, the column is composed of five styles glued together, and the five lobes are the stigmas. The hairs upon the ovaries will give you a good idea of the nature of hairs in other plants. Cut off, with a sharp knife, a very minute portion of the skin of one of the ovaries, and lay the morsel on its side in water under the microscope ; you will find that the hairs are delicate transparent projections, tapering to a point. Some of them are very short and

curved downwards, as if it was their business to
protect the surface of the ovary (*fig.* 11.), others
stand erect, and have a head of a brownish colour,
from which a clammy fluid exudes ; the last are
secreting hairs, and their duty is supposed to consist
in carrying off the volatile matter to which the plant
owes it smell ; for they are not only found on the
ovaries, but on almost all the other parts.

When the fruit (*fig.* 6.) is ripe, it resembles, in a
striking manner, the bill of certain birds ; on which
account the Geranium is called, in English, Cranes-
bill, by which I would have introduced it to you, if
the Latin name had not become the more common.
This singular appearance is owing to a very simple
circumstance. In most plants the styles shrink up,
or fall off, at the same time that the flower fades,
and by the time the fruit is ripe, have entirely disappeared. But in the Geranium, the styles continue
to grow and harden, as fast as the fruit itself ; and
when the latter is ripe, the styles project from
the ovaries in the form of a beak. At the time
when the fruit is ripe the seeds are shut up in the
cavities of the ovary, so that one would wonder how
they are to get out; if you would wish to catch the
Geranium in the act of sowing its seed, gather a
little branch of the ripe fruit in a fine summer's
morning, before the dew is off it, and put it in
the sun. By degrees the fruits will dry, and if you
watch them, you will be surprised by some of them,
on a sudden, emitting a snapping sound, and you
may see first one and then others of the ovaries

quickly curving upwards towards the top of the style, opening, at the same time, by their face, so as to let their seed drop out (*fig.* 6. *a.*). This is caused by the styles contracting from dryness, and shortening; they stick so close together at their points, that they cannot separate there, and so they actually pull the ovary up by the roots, and then roll up upon themselves, as if they were frightened at what they had done. The seeds are often beautiful objects, and are sometimes curiously pitted or netted all over their surface. No workman ever gave such finishing to the setting the most costly gems, as Nature has given even to the seeds of a weed.

Among our wild flowers are two genera which belong to the Geranium tribe; the Geranium itself, and the *Storksbill* (Erodium). The latter is very common in gravelly and sandy wastes, and looks so like a Geranium, that you will, most likely, mistake it for one. You may, however, know it by five only of its ten stamens bearing anthers. In the Geranium you will find that the five outermost of the stamens, are much shorter than those which form the inner row, and more stunted (*fig.* 2.), but they all have anthers. In the Storksbill, the five outer are still more stunted, and have no anthers. But the most interesting part of the tribe is that which is cultivated in green-houses, and to which I have already said the name *Pelargonium* is applied. This differs from both the other genera in the stamens being more than five, and fewer than ten in number, and in the corolla being irregular; that is to say, different in dif-

ferent parts; the two upper petals are larger than
the three lower, and stand altogether apart, so as to
give the flower the appearance of having two lips.
These Pelargoniums are almost entirely natives of
the Cape of Good Hope, and have become as much
the favourites of modern florists, as the tulip, the
pink, the ranunculus, and the auricula were of their
forefathers. And yet they were not originally so very
beautiful; their leaves indeed were always fragrant,
and their colours gay, but they possessed nothing
like the clearness of complexion, the regular features,
and the rich variety of colour which characterise
the present race. To what are we to attribute this
sudden change? To cultivation? No doubt to cultivation; but to cultivation of a very peculiar kind,
of which our grandfathers never dreamed; we call it
HYBRIDIZING.

You are already acquainted with the singular causes
which bring about the change of ovules into seeds
(see page 8.); and you know, that if the pollen does
not act upon the stigma, the ovules shrivel up and
die prematurely. It has been discovered, that if two
plants are very near relations, the pollen of one will
act upon the stigma of the other, just as well as if the
pollen was produced by the anthers of the plant to
which the stigma belongs; but when the seeds so
obtained are sown, they change to plants, which are
not exactly like either of those from the intermixture
of which they sprang, but which bear a strong resemblance to both. For instance, if you take the
pollen of a plant with blue flowers, and place it upon

the stigma of one which has red flowers, the seed will produce a plant having purple flowers; or, if a plant with a very vigorous mode of growth is thus intermixed with another of a very dwarf habit, the plants which spring from seeds thus procured, will be neither very dwarf nor very tall; and so on. This is the secret of the improvement of Pelargoniums, which happen to intermix very easily : a sort with large ugly flowers, is intermixed with one with small neat flowers, and you have, in all probability, a variety with large flowers, that are as neat in appearance as those of the small flowered kind. I need not particularize, with more minuteness, the way in which plants respectively influence each other ; a little reflection must render it apparent to you, now that you understand the principle. I must not, however, omit to tell you, that intermixture can only take place between plants very closely related to each other, and that distant relations have no influence the one on the other. You could not hybridize a Geranium with a Pelargonium, nor those Pelargoniums which have fleshy tumours for stems with such as have slender stems, nor even a gooseberry with a currant; but to the power of intermixing the slender stemmed real Pelargoniums there seems to be no limit.

That nature acted thus, long before man discovered her secret, there can be no doubt; for winds and insects are as skilful hybridizers as we are; and the different races of apples, pears, and other fruit, which have, in all ages, sprung up in gardens, are, no doubt, indebted for their origin to such circum stances ; but it is only in modern times that this mode of proceeding has been reduced within certain rules, and that hybridizing has become a fixed and useful art.

EXPLANATION OF PLATE II.

I. THE UMBELLIFEROUS TRIBE.—1. A portion of the flowering stem of *Fool's Parsley*, shewing the umbel ; *a a a* the one-sided involucres by which the plant is especially known.—2. A flower magnified, shewing the petals, stamens, styles, and disk.—3. A petal separated from the rest, with its curious incurved point.—4. A stamen, with the filament and anther.—5. A young ovary seen from the side ; *a* the ovary itself ; *b* the styles ; *c* the line where the calyx is to be sought for.—6. A ripe fruit viewed from its side ; at the base, a forked axis is just perceptible, from the arms of which the grains finally swing.—7. The fruit with its grains separated, and swinging from the arms of the axis.—8. A back view of one of the grains ; the five prominent ridges are distinctly visible.—9. The face or commissure of the same grain ; *b* one of the stripes or vittæ.—10. The same grain cut through ; *a* the ridges ; *b* the stripes ; *c* the albumen.—11. Half a seed taken out of the rind of the grain : *a* the embryo, lying at the upper end of a large quantity of albumen.—12. The embryo very highly magnified, and seen sideways for the sake of shewing its two seed-leaves, or cotyledons.

II. THE GERANIUM TRIBE.—1. A couple of flowers of *Meadow Geranium*, one of which has dropped its petals ; *a* bracts.—2. The ten stamens and the style.—3. The lower part of a pistil, covered with glandular hairs.—4. The top of the style and the five stigmas.—5. The lower part of a pistil split, to shew *a* the ovule, and *b* a tapering axis, to which the styles are glued. It is this part which forms the hard centre of the beak in the ripe fruit, and from which the styles drop, after they have curled up to shed the seeds.—6. A couple of ripe fruit, one of which is beginning to shed its seed ; *a* an ovary from which the seed has dropped.—7. A seed ; *a* the hilum, or scar left by its separating from the ovary.—8. Seed split ; the embryo is curiously rolled up, and plaited, and there is no albumen ; *a* is the radicle, or seed-root ; *b* the cotyledons, or seed-leaves.—9. Is a cross section of the same seed.—10. The embryo in its natural state, with the skin of the seed stripped off; the radicle is a taper body, hooked back at the narrow end of the embryo.—11. Hairs very highly magnified ; the short sharp-pointed ones are lymphatic hairs ; the long upright pin-headed ones are secreting hairs.

LETTER III.

(Plate III.)

My last letter was so long that I fear you must
consider me unreasonable in expecting you to be
already prepared for another. And yet I think that
considering the interest you say your little girls take
in plants, I shall scarcely do wrong in profiting by
the opportunity I happen to possess of going on with
the subject, especially as this is likely to be much
shorter. They already begin to wonder where the
terrible difficulties lie concealed, with which you have
hitherto been frightened from allowing them to study
Botany; but I can venture to assure them that, as
far as the elementary parts of the science are con-
cerned, there are no difficulties to encounter greater
than what they have already overcome. I feel perfectly sure that Crowfoots, and Poppies, and Umbel-
bearing plants, and Geraniums, are now familiar to
you, together with all their relations. We will next
take a family of quite another kind.

In the meadows and woods of Europe, North
America, and the colder parts of Asia, are found a
great number of herbs which, with a great accor-

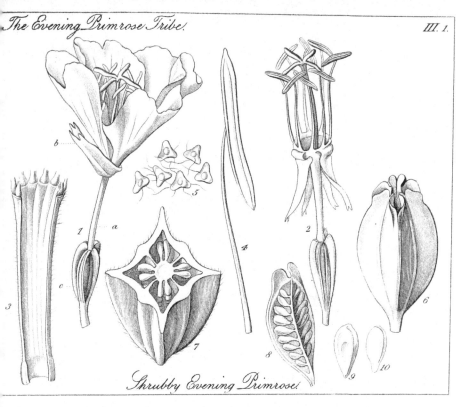

Shrubby Evening Primrose.

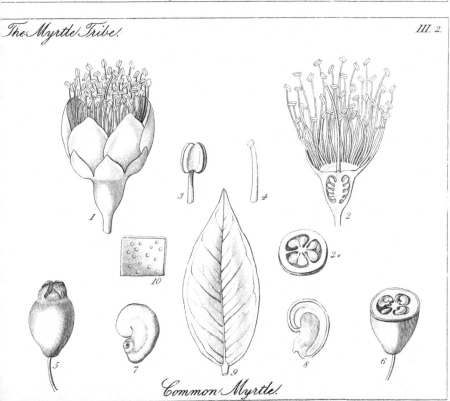

Common Myrtle.

dance in their general appearance, agree also in this remarkable circumstance, that every one of the parts of the flower consists either of four pieces or of some number which may be divided by four; in South America are many species of a similar nature, only they are shrubs and are much more richly coloured. Botanists call these Onagrariæ, or the *Evening Primrose tribe*, because the charming yellow flower which unfolds its bosom to the evening sun, and drinks up the dews of night with its petals, rendering darkness as lovely as noon-day, but which retires at the approach of the sun, rolling up its petals and carefully protecting its stamens and pistils from the glare of light, is one of the tribe; it might be called the owl of the vegetable world, only it is more beautiful and delicate than that hard-hearted enemy of mice.

If you have ever examined one of them accurately you will be at no loss to recognize all the rest. For this purpose suppose you take the *Shrubby Evening Primrose* (Œnothera fruticosa), a beautiful little North American plant, with an absurd name, for it is not a shrub.

The leaves of this plant are of a narrow figure, not unlike the head of a lance, and their veins are disposed in a netted manner like all the preceding; it has therefore a stem which increases in size by addition of matter to the outside of the wood; or in one word it is Exogenous; the leaves do not grow *opposite* each other from opposite sides of the stem, but are placed one a little above the other, so as to be *alternate*; mark this.

The flowers are of a bright yellow ; and are entirely different from those of any of the preceding tribes. In the first place the calyx has a long slender tube (Plate III. I. *fig.* 1. *a.*), from the top of which arise two leaves, both turned the same way, and notched at the point (*fig.* 1. *b.*); it is in reality composed of four sepals, united at the base into a tube, but capable of being separated above the tube into *four* pieces, as you may easily see if you attempt to divide it with the point of a pin.

From the top of the tube of the calyx arise *four* petals.—Observe, again,—four,—which are of a bright yellow, and are rolled together, except in the night, when the flowers are expanded.

Twice four stamens spring from the top of the tube; each has a very long anther, which swings by its middle from the summit of the filament, and sheds its pollen in such a way that it looks as if it were mixed with cobweb. If you magnify this pollen in a drop of water, each grain will be found three-cornered (*fig.* 5.), and held to its neighbour by excessively delicate threads; a peculiarity in the pollen which is rarely met with, except in the Evening Primrose tribe.

The ovary (*fig.* 1. *c.*) is inferior, and is marked by eight ribs, of which *four* are more prominent than the others; it contains *four* cavities, in each of which is a great many seeds. The style is a long slender body, rising within the tube of the calyx, as high as the stamens, and then separating into *four* narrow stigmas.

The fruit is a dry oval case, with *four* angles, opening into *four* pieces, called VALVES (*fig.* 6.).

Thus, you see all the parts of this plant, from its calyx to its fruit, consist either of four, or twice four parts ; the like happens in all the genuine species of the same natural order ; by which character they are easily known. There are many plants of very different orders, that have four sepals, or four petals, or some of their other parts, of that number; but it is only in the Evening Primrose tribe that all the parts are in fours at the same time ; or some multiple of four, which is botanically the same thing.

There are no Evening Primroses, really wild, in Great Britain, however frequent they may be in gardens. But there is an exceedingly common wild flower, called *Willow-Herb* (Epilobium), one of the species of which, called the " great hairy," is, perhaps, the most noble of all our British herbs. Its stout hairy stems grow five or six feet high, and are terminated by long clusters of bright red flowers. If you were to compare it with the description of the Evening Primrose, you would think it really must be a species of that genus, only the flowers are yellow. This, however, is not the only difference. When the fruit of the Willow-herb is ripe, it sheds seeds, which are furnished with a curious apparatus to enable them to fly about, and spread themselves over the land ; each of them has a very long tuft of silk at one end, which is so light, that the smallest breeze is sufficient to buoy it up, and raise it aloft into the air, there to be caught and carried to a great distance.

E

Nothing of this sort is found in the Evening Primrose.

Another plant, of far greater beauty than either of the foregoing, is the Fuchsia, an American genus, for which no English name has been contrived, and which is now one of the greatest of all the foreign ornaments with which our gardens are embellished in the summer and autumn. Every body has Fuchsias; the poor weaver grows them in his window; many an industrious cottager shews them as the pride of the little plot of ground before his door; and even the suburban inhabitants of London itself, speak of the beautiful Fuchsias they rear, with enthusiasm and delight. You must, therefore, know very well what the Fuchsia plant is. Examine its flowers; on the outside of all, you have a deep crimson covering, divided into four firm sharp-pointed leaves; this is the calyx. Rolled up within it, and closely embracing the stamens, are four little dark purple leaves, which are not half so long as the calyx; they are the petals. The other parts you will easily recognise. But the fruit is not a hard dry case, or capsule, bursting into four valves when it is ripe; it contains four cavities indeed; but its rind is deep purple, fleshy and juicy; in a word, it is a berry. This, then, is a marked distinction from other plants of the Evening Primrose tribe; but, as in all other respects the Fuchsia agrees with them, it is not accounted sufficiently different to belong to any other natural order.

The Evening Primrose tribe has little, except its beauty, to render it interesting to mankind; for there is not a single species which possesses any useful property worth mentioning Remember, that number four, throughout all the parts of the flower, is its character; and you will be in no danger of either forgetting it, or mistaking it.

I have already said, that other orders have, occasionally, four parts of the calyx, or corolla, or of some other class of organs, and yet do not belong to the Evening Primrose tribe. I will give you an instance of this.

You know what a Myrtle is. Take a sprig of that beautiful, but delicate evergreen, for your next subject. It has hard shining deep-green leaves, which do not drop off when winter comes; but seem as if they were intended to make us forget that winter has power over vegetation; they stand *opposite* each other, and if you bruise them, emit a fragrant aromatic odour. If you hold them against the light, you will see them look as if pierced with holes, closed up by a green transparent substance; they are not, however, pierced : but the appearance is owing to their containing a vast number of little transparent cells, in which the aromatic matter, to which they owe their fragrance, is laid up (Plate III. 1. *fig*. 10.).

The flowers have a calyx of five divisions; there are five petals of a dazzling white; and from the sides of the calyx, there arises, in a ring, a very considerable number of slender white filaments, tipped by little roundish anthers (*fig*. 2.).

The ovary, which is inferior, contains three cells, and a good many ovules; from its flat top springs

one style (*fig.* 2.), which ends in a stigma, so small
that it cannot be discovered without a microscope
(*fig.* 4.).

The fruit of the Myrtle is a purple berry, so like
a Fuchsia berry on the outside that you might mistake the one for the other ; but it has only three cells,
instead of four, and has a strong aromatic taste, of
which the Fuchsia is entirely destitute.

You will, after reading this, ask me, perhaps, with
surprise, what resemblance I can discover between
the Myrtle and the Evening Primrose tribe ; for it
seems difficult to select two objects more unlike.   I
answer thus—although the Myrtle itself is not very
like an Evening Primrose, yet there are many of the
Myrtle tribe, which, having only four divisions of the
calyx, and four petals, might be mistaken for plants
belonging to the Evening Primrose tribe, for they
have an inferior berry, like that of a Fuchsia ; you
would however see that the number four could
not be traced further than the petals, and, consequently, the resemblance would cease with these parts.

The Myrtle tribe, like the last natural order,
abounds in beautiful plants ; it also contains many
that are of great use. The spice you call *Cloves,* consists of the young flower-buds of a tree found in the
West Indies (Caryophyllus aromaticus) ; and *All-
spice* is the berries of another (Myrtus Pimenta). The
pleasant fruits called the Rose Apple, and the Jamro-
sade, in the East Indies, are produced by different
species of Eugenia ; Guava Jelly is prepared from the
succulent berries of trees of the Myrtle tribe, found

in the West Indies; and, finally, the Pomegranate
is an example of another fruit-bearing kind, which
has migrated from Barbary into Europe.

All these are kinds with berries for their fruit; and
they form the greatest part of the tribe. Others however there are which have dry fruits opening at the
top, and containing a great number of very minute
seeds; these, the principal part of which are natives
of New Holland, have very often also alternate leaves.
It is therefore neither to the fruit nor to the position
of the leaves upon the stem, that you are to look for
the precise character of the Myrtle tribe. *The inferior ovary, the numerous stamens, the single style, and
the dotted leaves*, are what you will know it by with
most certainty.

To that division of the tribe in which the fruit is dry
and many-seeded belong *Melaleuca* and *Metrosideros*,
with their long tassels of silken stamens, purple, or yellow, or crimson, and so do the gigantic *Gum Trees* of
New Holland (Eucalyptus). These last are remarkable
for having no petals; and for their calyx falling off like
a lid or extinguisher. I told you, in my first letter,
to observe what I said about the curious calyx of
Eschscholtzia, which was pushed off by the petals in
the form of a hutkin, in consequence of the sepals
not being capable of separating in the usual way. So
is it with the Eucalyptus; its calyx has all its parts
soldered together, as it were, into a hard fleshy lid;
when it is time for the stamens to unfold, they push
the calyx so forcibly, that it breaks away by its base
and drops off, leaving the stamens at liberty to expand as fully as may be necessary.

EXPLANATION OF PLATE III.

I. THE EVENING PRIMROSE TRIBE.—1. A flower of the *Shrubby Evening Primrose* half unfolded ; *a* the tube of the calyx ; *b* the divisions of the calyx ; *c* the ovary.—2. The same flower deprived of its petals, and with its sepals divided from each other, shewing the stamens and stigma.—3. The tube of the calyx cut open ; at the top are seen the bases of the eight stamens that spring out of it.—4. A stamen.—5. A cluster of pollen-grains, with the threads by which they are held together ; represented as they are seen when immersed in water.—6. A ripe fruit, with the four openings at the top, through which the seeds escape.—7. The ovary cut through, before it is ripe ; it exhibits the four cells, and the numerous seeds in them.—8. One of the valves of the fruit separated from the others, with the seeds sticking to it.—9. A seed.—10. An embryo, from which the seed-coat has been stripped ; it is not surrounded by albumen, but is protected only by the skin of the seed.

II. THE MYRTLE TRIBE.—1. A flower of the *Common Myrtle*, much magnified.—2. The same cut perpendicularly, shewing two of the cells of the ovary, and the origin of the style and stamens.—2*. A cross section of the same.—3. Another with part of a filament.—4. The tip of the style and stigma.—5. A ripe fruit.—6. The same cut across, with the curved embryo seen lying within it, without any albumen.—9. A leaf.—10. A portion of a leaf magnified, shewing the transparent dots in it.

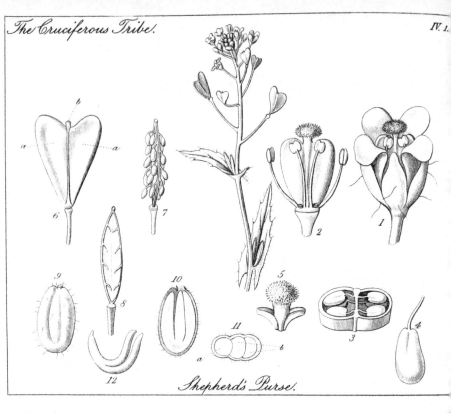

Shepherd's Purse.

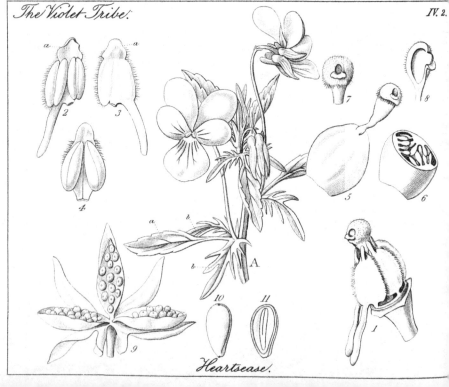

Heartsease.

LETTER IV.

THE CRUCIFEROUS TRIBE—DOUBLE FLOWERS—THE
VIOLET TRIBE—METHOD OF ANALYSIS.

(Plate IV.)

I AM pleased, but not surprised, to find from your
last letter, that Botany does not prove, upon examination, to be either so dry and technical a science as
you imagined, or so little related to objects of daily
interest, as others had persuaded you. Depend upon
this truth, before all others, that knowledge is always
useful, and that when we are unable to discover its
utility, we have to blame our own short-sightedness,
or incapacity, or any other cause, rather than knowledge itself.

Sir John Herschel, in that admirable treatise of his
which should be in the hands of every human being
who has arrived at an age to be capable of under-
standing it,—Sir John Herschel has well remarked
that, "the question to what practical end and advantage do your researches tend? is one which the spe-
culative philosopher, who loves knowledge for its
own sake, and enjoys, as a rational being should
enjoy, the mere contemplation of harmonious and
mutually dependent truths, can seldom hear without
a sense of humiliation. He feels that there is a lofty

and disinterested pleasure in his speculations which
ought to exempt them from such questioning ; com-
municating as they do, to his own mind the purest
happiness (after the exercise of the benevolent and
moral feelings), of which nature is susceptible, and
tending to the injury of no one ; he might surely
allege this as a sufficient and direct reply to those
who, having themselves little capacity, and less re-
lish for intellectual pursuits, are constantly repeating
upon him this inquiry. But if he can bring himself
to descend from this high but fair ground, and justify
himself, his pursuits, and his pleasures in the eyes of
those around him, he has only to point to the history
of all science, where speculations, apparently the
most unprofitable, have almost invariably been those
from which the greatest practical applications have
emanated. What, for instance, could be apparently
more unprofitable than the dry speculations of the
ancient geometers on the properties of the conic
sections, or than the dreams of Kepler (as they would
naturally appear to his contemporaries) about the
numerical harmonies of the universe. Yet these are
the steps by which we have risen to a knowledge of
the elliptic motions of the planets, and the law of
gravitation, with all its splendid theoretical consequences, and its inestimable practical results. The
ridicule attached to ' *Swing-swangs*' in Hooke's time
did not prevent him from reviving the proposal of the
pendulum as a standard of measure, since so admirably wrought into practice by the genius and perseverance of Captain Kater ;—nor did that which

Boyle encountered in his researches on the elasticity
and pressure of the air, act as any obstacle to the
train of discovery which terminated in the steam-
engine.   The dreams of the alchemists led them on
in the path of experiment, and drew attention to the
wonders of chemistry, while they brought their advo-
cates (it must be admitted) to merited contempt and
ruin.   But in this case it was moral dereliction
which gave to ridicule a weight and power not necessarily or naturally belonging to it : but among
the alchemists were men of superior minds, who
reasoned while they worked, and who, not content
to grope always in the dark, and blunder on their
subject, sought carefully in the observed nature of
their agents for guides in their pursuits ;—to these
we owe the creation of experimental philosophy."

It perhaps would not have been amiss to have
begun this correspondence with the quotation I have
just made ; but I was anxious that a little interest
should be awakened in the subject before any thing
like a formal defence of its utility should be under-
taken by me.   If I were disposed to add any Botanical instance of important results arising from apparently trifling causes, it would be easy enough to do
so.   For instance, the microscopical investigations by
Grew of the nature and properties of the little cells
and bladders that are found beneath the skin of
plants, were the forerunners of all the valuable train
of physiological discoveries, by which the productive-
ness of the soil has been so much increased; the curious but neglected inquiries of Kölreuter into the

possibility of intermixing the races of plants, laid
the foundation of modern improvements in the qualities of cultivated species; while the ingenious but
derided speculations of Camerarius upon the relation
that exists between the properties and structure of
species, have put the physician in possession of a
power of discovering the hidden uses of plants, the
limits to the application of which no one can
foresee.

It is mentioned, that in the voyage of Lord Anson
round the world, when new and unknown lands were
constantly discovered, the dread which his surgeon
entertained of the effect of strange herbs was so great
that, from fear of poisoning the crews, he would sometimes permit them to use no other kind of fresh vegetable food than grass. At the present day there
should be no navy surgeon who would not be able to
point out at once, in every place, an abundance of
plants, the use of which could not by possibility be attended by any ill effects. You have already seen that
the Crowfoot tribe consists of burning and blistering
species, that the Poppy tribe produces stupefaction,
the Umbelliferous tribe is chiefly aromatic, but not
always to be trusted, Geraniums astringent, Evening
Primroses insipid, and Myrtles fragrant and aromatic.
Another example of the uniform prevalence of peculiar properties in the same tribe or natural order,
is afforded by Cruciferous plants.

The healthy stimulating effects of Mustard and
Cress, and the nutritive properties of Turnips and
Cabbages are well known to every body. These

plants belong to an extensive tribe called Cruciferous, or Cross-bearers, because their four petals are placed in such a way as to resemble in some degree a Maltese cross. Unfortunately for those who have little power of observation, and less patience, their flowers are usually very small ; but I am convinced that this circumstance will be far from deterring my young friends from attempting their study ; it will rather operate as an incentive to their making themselves acquainted with them.

They are already, I suspect, familiar with a mean-looking weed, called *Shepherd's Purse* (Capsella Bursa pastoris), which is found every where at all seasons of the year, except the severest part of the winter. Its name was given it because it has a number of pouches filled with very small seeds, which you might fancy were fairy coins. Let us look at it botanically.

Its leaves are veined in that netted manner which indicates the Exogenous structure; and consequently you know that if it were ever to form a woody stem, its woody matter would be arranged in concentric circles. The form of the leaves is like that of an ancient arrow head, sitting closely to the stem, and extended downwards at the base into a sharp barb on either side (Plate IV. 1.).

The flowers are arranged regularly upon a central stalk in the form of a *raceme;* and, what is extremely singular, they are uniformly destitute of bracts. This is so unusual a case that I do not remember any other instance in the whole vegetable kingdom in which bracts are constantly absent ; the absence of these little leaves is hence a mark of the Cruciferous tribe. Observe, I pray you, how very useful it is to be aware of this. Imagine yourself cast away upon a desert island; and there, surrounded by plants of unknown forms and tempting looks, none of which you dare use from fear of their proving poisonous. Among them however you remark a good many of the same kind, one of which is just beginning to bear its tufts of flowers : the blossoms are too young to be examined, but old enough to shew you that they grow without bracts ; the leaves you would easily see were those of Exogenous plants, and you would immediately know that this species at least would be not only harmless, but the very best kind of vegetable for you to consume ; a salad which might be eaten with the utmost confidence.

But it is not thus alone that Cruciferous plants may be recognized. The structure of their flowers is of a very peculiar kind. The calyx is formed of four little leaves or sepals ; within which are four very small white petals, arranged in the manner which I have already stated gave rise to the appellation of Cross-bearers (*fig.* 1.). Within the petals are six stamens (*fig.* 2.), of which two are a very little shorter and more spreading than the other four. To this character no parallel is to be found in any other than Cruciferous plants, and consequently it is a second essential character, by which they are to be known.

The pistil is an oval green body, shaped something like a wedge, on the summit of which is a little cushion of a stigma seated on an exceedingly short style (*fig.* 2.). If you cut open the ovary you will find (*fig.* 3.) that it contains two cells, in each of which is a number of young seeds or ovules hanging by slender thread-like stalks.

The fruit (*fig.* 6.) becomes a wedge-shaped flat body, composed of three pieces, two of which (*fig.* 6. *a. a.*), the valves, separate from the third (*fig.* 6. *b.*), which is named the *partition* or DISSEPIMENT (*fig.* 8.) ; it is to the edges of this third piece that the seeds stick by little threads (*fig.* 7.). In the inside of these seeds the embryo is bent double, after a singular fashion (*fig.* 10.), the seed-root being pressed close to the back of the seed-leaves.

This has been rather a wearisome lesson, to one so little accustomed to the use of the microscope, to which you must already have had recourse several times ; but you have now the satisfaction of knowing that you possess the secret of recognizing with certainty nearly a thousand species, scattered over the face of the world, all of which are harmless, and many highly useful.

What would the farmer do without Turnips and Rape, which are Cruciferous plants ? or the gardener without Cabbages, Sea-Kail, Mustard, Cress, and Radishes? or how could the florist supply the place of his Wall-flowers, and Stocks ? All these are such common plants, that you can have no difficulty in procuring specimens for examination ; you will find that while they are all unlike in trifling circumstances, they agree in having their parts arranged exactly in the same manner as the Shepherd's Purse;

but you will remark a difference in their fruit, of this
nature ; in some of them, as the Shepherd's Purse
itself, the pod is so very short, that there is not
much difference between its length and breadth ; in
others, it is very long and slender, as in the Turnip
and Cabbage; to the former of these fruits the name
of *Silicle* is given; to the latter, that of *Silique;* and
by many Botanists the whole tribe of Cruciferous
plants is divided into two portions, of which one is
called *Siliculose*, and the other *Siliquose*. By the more
scientific Botanist of the present day, this distinction
is held in less estimation, and new divisions founded
upon the structure of the embryo, are employed ;
but these are so obscure, that I will not fatigue you
with them.

The most beautiful species in the whole tribe are
the Wall-flower, which sheds its sweetest odours over
the ruined buildings of England, and the Stock,
that, with its hoary leaves and gay flowers, gives an
air of green old age to the rocks and cliffs of the
Mediterranean. Both, these, however lovely in their
wild and single state, are chiefly cultivated when
their flowers have become what is called double ;
that is to say, when the parts which are usually stamens, and pistils, and sepals, are all transformed
into petals; by which means the quantity of gaily
coloured parts is much augmented. There are those
who, with an air of scientific delicacy, pretend to
despise these beautiful objects, and call them mon-
sters ; but I would not have you follow such an
example ; for, in reality, double flowers, indepen-

dently of their acknowledged beauty, afford the most important evidence of the true nature of the different parts of the floral system, as you may one day know. To you, as to many others, it may be a subject of wonder how these double flowers are increased, for if the stamens and pistils are converted into petals, it would seem that no means are left for multiplying the race. This would, doubtless, be so, if all the stamens and pistils were really thus transformed ; but among many flowers, some are found in which a perfect stamen or two remain; and others, in which a perfect pistil or two can be found. If the stigmas of the latter are touched with the pollen of the former, ovules are fertilized, and seeds produced, which will grow into other plants, the flowers of which will be as double as those of their parents. No one knows why double flowers should be capable of being thus perpetuated ; it seems as if any tendency which is once given to a plant, may be carried on from generation to generation, by a careful attention to the stoppage of all disposition to depart from the new character : in the Stock any plant that produces flowers less double than usual, must have a tendency to depart from the double state, and, therefore, should not be allowed to bear seed, or to influence the seeds borne by other plants, but should be carefully eradicated as soon as its flowers are sufficiently expanded for their true character to be ascertained. By attention to such rules, Turnip-rooted and Long-rooted, White, and Scarlet, and Purple Radishes, and all the different races of Turnips, have been preserved for years; whereas, if great precautions to maintain them in their purity had not been constantly taken, they would long since have become thrown together, and reconverted into the wild form from which they sprang. I take it for granted, you are so much interested in the pretty flowers of your garden, that you would be sorry to see all the double ones turn single:—you will now be able to avert such a sad catastrophe.

Candy tuft (Iberis), sweet *Alyssum*, the snowy *Arabis* of spring, that pretty little tufted purple thing which is named after the French flower-painter Aubriet (Aubrietia deltoidea), *Honesty* with its clusters of broad bucklers, the modest *Whitlow-grass* (Erophila verna), which springs up on the crest of every wall, the earliest harbinger of spring, *Watercresses*, *Horse Radish*, and a host of others, will be furnished either by the garden, or the fields, to augment your acquaintance with this natural order. Leaving it now, as requiring no further explanation, we will next proceed to another spring tribe.

Violets, sweet Violets, and Pansies or *Heartsease*, represent a small family (Pl. IV. 2.), with the structure of which you should be familiar; more, however, for the sake of its singularity, than for its extent or importance; for the family is a very small one, and there are but few species belonging to it in which much interest is taken. As the parts of the Heartsease are larger than those of the Violet, let us select the former in preference, for the subject of our study.

The Heartsease is a little herbaceous plant, as you

know, with leaves cut, as you suppose, into several deep divisions; here, however, you are mistaken. The true leaf is a narrow oblong blade, with netted veins, rather notched at its edge, and tapering gradually into a stalk (*fig. A. a.*); it is not slashed or divided at all. But on each side of the leaf, quite at the bottom of its stalk, there is a deeply-cut stalkless part, which is of the same colour as the leaf, but shorter (*fig. A. b. b.*); these are stipules (see page 37), and it is they which give the lobed and lacerated appearance to the leaf. Here you have an exemplification of the care with which you ought to look at plants if you would understand their construction rightly; it may be true indeed, as theoretical Botanists say, that stipules are only little leaves, but it does not follow that on that account we may call them leaves; for it is quite clear that whatever their theoretical similarity may be, they are stationed by nature in a particular place, in a particular form, for some good purpose or other. You will some day know that the sepals of the calyx, and the petals of the corolla, and the stamens, and the carpels, are all leaves in different states; but you must not on that account cease to distinguish them carefully, and call them by their right names, when you find them fixed by nature in the form of sepals, petals, stamens and carpels. Our own foot is a sort of hand, and our toes are fingers; but we cannot on that account dispense in practice with the use of the words feet and toes.

The flowers consist of five sepals of a narrow

F

figure and extended in a singular manner at the base; of these some are much larger than the others.

The corolla is formed of five petals, which are also of unequal size; two of them, which are differently coloured, stand erect and rather above the others; a third, standing in front of the rest, has a short horn or spur at its base.

Then we have the stamens, also five in number, of a singularly irregular form (*fig.* 1.); two of them, which are in front of the others, have long tails, that are hidden within the horn of the 'front petal; the other three have no tail, nor any particular irregularity of figure, but they are all terminated by a broad membrane of a rounded form (*figs.* 2. 3. *a. a.*), and bordered by a fringe of hairs; filaments there are none.

The pistil is a superior roundish pale-green body (*fig.* 5.), terminated by a short fleshy style, which is shaped like a narrow funnel, or a taper inverted cone; at the top (*figs.* 7. 8.) it is of a bright green, nearly spherical, slightly hairy, and hollow, with a hole on one side, to which there is a minute lip; through this hole there is access to the stigma; no one has yet discovered for what purpose such a singular conformation has been provided. The ovary contains but one cell; but, as in the Poppies, it has three projecting lines running up its shell at equal distances in the inside, and covered with young seeds (*fig.* 6.).

When the fruit is ripe it is still surrounded by the calyx (*fig.* 9.), although both petals and stamens have dropped off; it is an oblong shining case, which splits into three pieces or valves, in the middle of each of which stick the pale chesnut-coloured shining seeds.

There is no material difference from this structure in such other plants of the Violet tribe as you are ever likely to meet with; and therefore I shall suppose that the whole may be recognized by the characters I have explained.

Beautiful as they all are to look at, they would produce anything rather than Heart's ease if you were to eat them : for their roots have the property of producing sickness in so powerful a manner, that they are sometimes used in medicine as emetics. I would therefore advise you to confine your admiration to their beauty or their fragrance.

The Sweet Violet will have no rival among flowers, if we merely seek for delicate fragrance, but her sister, the Heartsease, who is destitute of all sweetness, far surpasses her in rich dresses and gaudy colours. She has become of late a special favourite with florists, who cultivate I know not how many distinct varieties, some of which have flowers of yellow and purple, or all yellow, or all purple, or nearly white, with every gradation of tint and depth, which one can well imagine. Methinks, I hear my young friends exclaim, are these fine plants, indeed, our humble Pansy, changed by cultivation? is it possible that the little drooping weed, which we have so often gathered among the stubble of cornfields in the autumn, can ever become the gaudy flower of the florist? Even so indeed is it; the

savage woad-stained Britons, were not more different
from the well-dressed ladies of the present day, than is
the Heartsease from its wild state, since it has attracted
the notice of the gardener. Those children of the
wild Pansy, which you see in the borders of the flower-
garden, have intermarried with strangers from other
climates, and especially with one from the Altaic
mountains (Viola altaica), where the race is finer and
more vigorous than beneath our northern sky.

Before I close this letter let me survey the ground
we have passed over. Eight distinct Natural Orders
have been examined, all of which are so very easily
known from each other, that it is almost superfluous
to repeat their characters ; yet as there is no more
certain method of fixing these matters in the memory
than by recapitulation from time to time, I must not
only do so, but beg of you to endeavour to get the
distinctive characters clearly understood and remembered by our little students.

Let me now vary the mode of distinguishing them,
and put their characters before you in a new form.

Of the eight orders, three have an inferior ovary,
and five have a superior ovary ; we will take this very
conspicuous character, as a preliminary mode of
distinction ; for thus we shall simplify the other differences. Then, of the three which have the inferior
ovary, the Evening Primrose tribe has its parts of
fructification in fours, the Umbelliferous tribe bears
flowers in umbels, and the Myrtle tribe has a great
many stamens and aromatic dotted leaves. On the

other hand, those with superior ovaries may be first separated into such as have a great many stamens, and such as have only a small and certain number. Of the former you have the Crowfoot and the Poppy tribe, the first of which has several distinct carpels, and the last all the carpels grown together into a hollow case ; of the latter the Geranium tribe has the carpels separating from a long hard beak-like centre, the Cruciferous tribe has six stamens of which four are longer than the others, the Violet tribe has anthers with a membranous crest, and a fruit splitting into three valves, to the middle of which the seeds are fastened. This will be still clearer to you if put into the shape of a table.

Ovary inferior
- Parts in fours—*Evening Primroses.*
- Parts not in fours
  - Flowers in umbels—*Umbelliferous Plants.*
  - Flowers not in umbels—*Myrtles.*

Ovary superior
- a great many stamens
  - carpels separate—*Crowfoots.*
  - carpels joined into a hollow case } *Poppies.*
- few stamens
  - carpels separating from a long beak } *Geraniums.*
  - carpels not separating from a long beak
    - Five stamens—*Violets.*
    - Six stamens, two of which are short } *Cruciferous Plants.*

This table is of no other use than to shew you how to analyse the characters of the subjects you examine; it does not give you, as you must remark, a correct notion of the *essential* characters of any of the tribes, but it states clearly how they *differ* from each other. They differ from each other in many other respects, but it was not necessary to express any thing further in order to enable you to know them from one another.

EXPLANATION OF PLATE IV.

I. The Cruciferous Tribe.—A sprig of *Shepherd's Purse.*— 1. A flower with all its parts in their natural position.—2. The same flower without the calyx and corolla; it shews the two side stamens, which are the shortest.—3. An ovary cut across, exposing the partition, the two cavities and the young seeds, or ovules.—4. An ovule apart, with the end by which it hangs from the side of the partition.—5. The stigma, with the style and a portion of the shoulders of the ovary.— 6. A ripe silicle; *a a* the valves; *b* the point of the partition.—7. The partition, from which the valves have been removed, shewing the numerous seeds which hang to it.—8. The partition seen in front, with the marks of the places to which the seeds were attached.—9. A ripe seed, covered with fine hairs.—10. The same cut through perpendicularly, shewing how the embryo is doubled up within it.—11. The same seed, cut through horizontally; *a* the radicle; *b* the two cotyledons.—12. An embryo pulled out of the seed-coat and straightened.

II. The Violet Tribe.—A. A piece of *Heartsease;* *a* blade of the leaf; *b b* stipules.—1. A view of the stamens and pistil in their natural position, after the petals and sepals are pulled off.—2. One of the horned stamens seen in face; and, 3. the same seen from behind; in both *a* represents the membrane, which is characteristic of the natural order.—4. A regular anther, seen in face.—5. A pistil.—6. The ovary cut through to exhibit the three elevated lines on which the seeds are placed.—7. A front view of the hole that leads to the stigma. —8. The head of the stigma split open, with a view of the stigmatic interior.—9. A ripe fruit split into its three valves.—10. A seed.— 11. The same cut perpendicularly to shew the embryo lying in the midst of albumen.

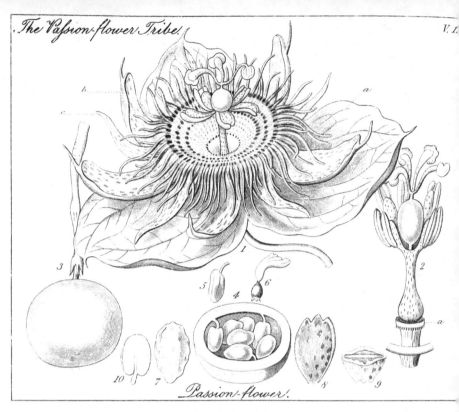

Passion-flower.

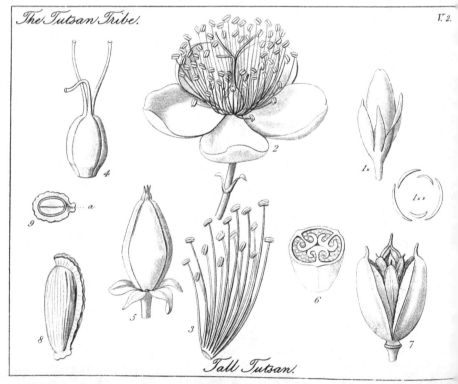

Tall Tutsan.

LETTER V.

(Plate V.)

WHEN the Spaniards discovered America they found, among other curious things, a flower, which they thought was an allegorical representation of the crucifixion and sufferings of our Saviour. In its anthers they saw his five wounds, in the three styles the nails by which he was fixed to the cross, and in a column which rises from the bottom of the flower the pillar to which he was bound; a number of little fleshy threads which spread from its cup, they compared to the crown of thorns. There are cuts, says, Sir James Smith, to be found in some old books, apparently drawn from description, like the hog in armour upon our signs to represent the rhinoceros, in which the flower is made up of the very things themselves (*Rees's Cyclopædia*).

Such travellers' stories as this, would now find few persons credulous enough to believe them; the tale is, however, not wholly fabulous. Like many others of the same sort it is made up of truth mingled either with falsehood or excessive exaggeration. There is such a flower and a highly curious one it is; and so far from rare, that you are probably already well acquainted with it, for it is now frequently to be seen trained even to cottage walls.

You are surprised at this, and I dare say wonder what this strange flower can be, that you have so often seen without discovering in it any of the marvels of the good Spaniards. They called it, in allusion to its mystical attributes, *Flos Passionis*, a Latin name signifying—Passion-flower, which the moderns have retained.

Many species with this remarkable character are known in gardens; they are all Passion-flowers, and are the representatives of the *Passion-flower tribe.* In other countries, other singular plants are also found, not exactly Passion-flowers, but belonging to the same tribe ; as you are never likely to meet with them, I need not trouble you about the manner in which they are distinguished. We will satisfy ourselves with a botanical view of that to which so strange a story is attached.

The Passion-flower is a twining plant, which helps itself to rise upon others by tendrils like those of the Pea, with which it often scrambles to the tops of high trees, or, if it misses its hold in the ascent, or is by any accident separated from the prop it has selected, hangs down among the branches in elegant festoons. Its leaves are veined in the netted manner ; and are often divided into deep lobes, but not always; their stalks bear here and there upon their upper edge little hard dark-green shining warts, called *glands ;* and they have a pair of stipules at their base.

The stems when cut through, shew the Exogenous structure ; a circumstance you would have known by aid of the leaves alone, if you had not had the stem to cut.

The flowers have on the outside three large bracts (Plate V. 1. *fig.* 1. *a.*), which together form an involucre.   Next these is a calyx, composed of five sepals (*fig.* 1. 6.) that are generally green on the outside, and differently coloured in the inside ;   sometimes blue, sometimes purple, now and then yellow or some other colour.   Let me particularly call your attention to this, which is a proof that the calyx, in other plants as well as in the Fuchsia, is occasionally coloured like petals.   When you are more of a Botanist you will find that this fact is connected with a curious tale of *vegetable transformations*, which I may some day relate to you.   At the base the sepals are joined together into a shallow cup, from which the petals and other parts arise.

The petals (*fig.* 1. *c.*) are always of the same colour as the inside of the sepals; but are nearly alike on both sides, are narrower, and are without a singular little horn, which projects from the back of the sepals, just as other horns spring from the corslet of certain beetles.

Next the petals come—the stamens, you will say :—not at all.   Next the petals come several rings of beautiful fleshy threads, which spread from the cup like rays, and are splendidly mottled with azure and crimson and white.   If there be one part of a plant more beautiful than all others, it is this

ray (or crown of thorns, as the Spaniards called it)
in the Passion-flower; the crimson blotches upon
it really do look like stains of blood. It diminishes
gradually in size towards the inside of the flower,
till at last it loses itself in some little rings, one of
which often surrounds the base of the column we
have still to examine in the centre (*fig.* 2. *a.*). Botanists themselves are hardly agreed upon the real nature of these singular rays; while some think them
imperfect petals; others think them imperfect stamens: a question of very little moment, and which
you are, as yet, unprepared to discuss; they are,
probably, parts in a state of change from one to the
other.

In the very centre of the flower, from the bottom
of the cup, rises a column (*fig.* 2.), at the top of
which are five stamens, each with a narrow two-
lobed anther, swinging from the point of a flat filament; you will wonder to see that these anthers, instead of turning their faces to the stigmas, like most
other anthers, are so contrived as to turn their backs
upon them; so that when they burst, the pollen cannot
fall upon the stigmas. This, however incomprehen-
sible an arrangement it may appear to us, is by no
means an event of unfrequent occurrence, as you will
hereafter discover; nobody has yet found out the cause
of it. What now is the column from which the stamens
seem to arise? The base of the filament, you say. To
a certain extent, you are right, The outside of the
column, which is speckled like the filaments, is constructed as you imagine; but if you cut it, you will

find it is only a hollow sheath surrounding a solid slender cylinder, from the top of which grows the ovary. It is an unusual thing to find ovaries that have a stalk in the inside of the flower : but such a structure is not found in the Passion-flower alone.

The ovary is an egg-shaped part, which, when you cut it, is seen to consist of a single cavity, from three elevated lines in which spring the ovules, just as in the Violet. It is surmounted by three styles, which are thicker at top than at bottom, and terminated by thick swoln stigmas. This peculiar form, no doubt, suggested the idea of their representing the three nails of the cross.

Thus you see, when this flower is stripped of all that is fabulous, there is still enough left in it to excite our admiration.

The fruit is, in all cases, a fleshy egg-shaped body, containing a number of pulpy seeds ; but it varies exceedingly in size and colour in different species. In the common blue Passion-flower, it is about as big as a hen's egg, and orange-yellow ; in others, it is smaller, and quite round (*fig.* 3.); in others, it is as large as a child's head. Fruit of the latter size are sometimes cultivated for the sake of the sub-acid pulp they contain, and are called *Granadillas* ; they are more esteemed in tropical countries, where the eatable fruits are generally bad, than in England, where we have so many really delicious fruits of our own.

If you were to look at the seed, you would say it was merely covered by pulp ; there is, how-

ever, something even here, beyond what you at first
perceive. Were you to watch the ovule in its progress to become a seed, you would remark a fleshy
sort of jacket, gradually rising from the bottom of the
ovule, overspreading its surface day after day, till it
had completely enclosed it : and then, on a sudden,
becoming soft and pulpy. Such a part as this we
Botanists call the ARILLUS; a part you have often seen
in another plant without knowing what it was. The
spice called *Mace*, overspreads the Nutmeg, as its
jacket does the seed of the Passion-flower, and is the
arillus of that aromatic production.

The seed may be easily deprived of the pulpy
arillus, which will strip back (*fig.* 6.); and then you
will discover that the seed itself is a blackish body,
with a brittle sculptured shell (*fig.* 8.). I shall not
trouble you about the contents of the seed, further
than to say, that they are sweet like a nut, and as
good to eat.

The only plant belonging to the Passion-flower
tribe, which you will find in the gardens, besides
Passion-flower herself, is a genus named *Tacsonia*
(see *Botanical Register*, tab. 1536.), which is also
found in South America, and is so like a Passion-
flower, that you will hardly distinguish it, except by
the very long tube of its flower. Its rays are short,
so that it has, in some respects, less beauty ; but
the richness of its colours, and the large size of all
its parts amply compensate for this defect.

But, although there is no other genus of the same
tribe within your reach, there are several belonging to

another tribe, with which your children must be familiar, without calling science to their aid. They certainly
have practical Botany enough to know a Melon, and a
Cucumber; and, probably, also a Gourd, a Vegetable
Marrow, and a Spirting Cucumber. Those things form
part of a small natural order called after one of them,
the *Gourd Tribe*, which is in near *affinity*, as we say,
with the Passion-flowers. Now, this word affinity is
one of the great practical difficulties in the way of
the student of the Natural System of Botany. Not
that it need be made so; for I have no doubt I could
have taken you to the end of our intended journey,
without saying a word to you about the matter. But
to understand it is essential to those who would form a
higher notion of Botany than what can be gathered
from the mere power of distinguishing one thing from
another; and if it could be comprehended, would form
a great aid to you in your future progress. I shall,
therefore, take the present opportunity of saying
something about the way in which the word is applied, in the hope that I may make it clearer to you
than it seems to be to many. Nevertheless, if you
do not understand me, you may skip all that follows
upon the subject.

AFFINITY signifies resemblance in *most* characters
of importance. It differs in degree just as resemblances between animals; which you can see and
understand more readily than those between plants.
Thus a monkey or baboon are very nearly related to
man, although totally distinct; that is, they resemble man in *most* characters of importance; and
are therefore in affinity with him. Again, a cat and

a lion agree in a very great number of their principal points of organization; and are therefore in affinity also. But a cat and a bird, although both of the animal kingdom, disagree in the greater part of their structure; they therefore are not in affinity with each other.

To take an illustration from plants you are now familiar with, compare the Crowfoots with the Myrtles; they both bear flowers composed of calyx, corolla, numerous stamens and pistils, and they both have leaves with netted veins and consequently an exogenous structure; but here their resemblance ceases. Compared as to other circumstances they are extremely different; as you will see by studying the two following columns :—

CROWFOOTS have	MYRTLES have
Lobed *leaves*, with an acrid watery juice, and usually with an alternate insertion on the stem.	*Leaves* not at all lobed, usually with an opposite insertion on the stem, and with a volatile oily juice, which is lodged in little transparent spots.
Numerous *stamens*, arising from below the carpels.	Numerous *stamens* arising from the sides of the calyx.
A superior *pistil*, consisting of several carpels, either not at all or but slightly adhering to each other.	An inferior *pistil*, consisting of several carpels, which are all grown into one solid body at the top of the fruit-stalk.
As many *styles* as carpels.	Only one *style*, whatever the number of carpels.
A very little *embryo*, which is furnished with a great quantity of albumen for its nourishment when young.	An *embryo*, which is supplied with no albumen for its nourishment when young.
They are chiefly *herbs*.	They are all *trees* or *shrubs*.

Thus in all these six important circumstances the Crowfoots and Myrtles are extremely dissimilar; therefore they are *not* in affinity with each other.

Next let us compare Evening Primroses and Myrtles, which we have seen disagree in the former having the parts of the flower always divided by four, no oil spots in their leaves, and but a small number of stamens. Compared as to other circumstances they are extremely similar, as the following columns will shew :—

Evening Primroses have	Myrtles have
Leaves sometimes opposite.	*Leaves* usually opposite.
Stamens arising from the sides of the calyx.	*Stamens* arising from the sides of the calyx.
An inferior *pistil* with many seeds.	An inferior *pistil* with many seeds.
Only one *style*.	Only one *style.*
A *fruit* which is sometimes pulpy.	A *fruit* which is usually pulpy.
An *embryo* with no albuminous provision for its infancy.	An *embryo* with no albuminous provision for its infancy.

Thus in these important circumstances the Evening Primroses and Myrtles essentially agree ; therefore they *are* in affinity with each other. You will further remark, that the points of difference between them in structure are no greater than what I just now mentioned ; so that the points of resemblance *are much more numerous than the points of difference.*

Having thus given you an idea of the meaning of the word affinity as used in Botany. Let me resume my account of the Gourds which are in near affinity with the Passion-flowers.

A Cucumber, which is one of the Gourd tribe, has a twining, scrambling stem, and raises itself

by tendrils ; its leaves have the netted arrangement of veins ; its flowers have a calyx which is like
petals in colour ; its stamens grow into a central column ; its ovary has only one cavity, with
the seeds attached to three lines which pass up its
sides ; its ripe fruit is succulent in the inside ; and
its seeds have a sweet nutty taste. All this reads as
if I were really talking of a Passion-flower; it is
these numerous points of resemblance, in important
points of structure, which constitute the affinity
between the tribes of Passion-flowers and Gourds.
In other respects they are materially different.

The Cucumber has very rough leaves ; it has no
petals ; its stamens grow in one flower, and its pistil
in another ; the ovary is inferior ; and there is no
trace of the beautiful rays of the Passion-flower. All
the Gourd tribe participate in these differences, which
thus become the essential characters of that natural
order.

The Passion-flowers are all harmless, and the fruit
of many of them is eaten. Thus we have another
similarity you will exclaim. Not quite so fast, if
you please ; I would not advise you to adopt that
idea practically, for if you do you may share the
fate of the poor sailor, who lately perished, as the
newspapers tell us, from drinking out of a gourd-
shell. In some countries there are Gourds with a
very singular figure ; they resemble a Florence flask,
such as oil comes home in, and have a hard rind
filled with soft pulp. Very useful bottles are prepared from such fruit, by cutting off the end of the

narrow part which represents the neck of the bottle, and then scooping out all the inside ; but it is necessary before using them that all the pulp should be removed, and that water should be allowed to stand in them, and be changed several times, till all the bitterness in which the rind abounds be removed ; owing to the purpose to which the fruit is then applied, the plant itself is called the Bottle-Gourd.

The bitter matter which is thus removed by washing, is not only unpleasant, but actually poisonous, as the unfortunate accident I have alluded to sufficiently proved. You will be surprised to hear that it exists universally in the whole of the Gourd tribe, even in the Cucumbers and Melons which you have so often eaten without being poisoned ; the truth is, that in these fruits the bitterness is dispersed through so large a quantity of pulp, and there is so little of it, that we are not sensible of its presence ; while in the Bottle-Gourd and others, it is so highly concentrated as to become dangerous. That it is found even in the Cucumber you may easily believe, if you call to mind how often that fruit is bitter even when upon the dinner table.

You will therefore recollect that the Passion-flower tribe is universally harmless ; but that the Gourd tribe is so often unwholesome, that the two or three instances you know of its fruit being eatable are to be considered exceptions to the rule.

The length of this letter has already so much exceeded my intention, that I must bring it to a close

with a brief account of a little natural order of wild
flowers, which we may dismiss without giving you
much more to learn.

The old herbalists had a plant which they called
Tutsan (Hypericum); a corruption of the French Toute-
sain, which we might translate " Allheal" ; they also
named it Androsæmum, which being translated signifies " Mans-blood," an odd name, that originated in
the soft fruit staining the fingers red when bruised,
and in a deep red colour being communicated by the
leaves to oily or spirituous medicines in which the
plant was often employed.   This and others of a
similar kind are common in meadows, and bogs,
on heaths, in groves and thickets, and by way-sides,
which they adorn with their bright yellow flowers.
The species which I have selected for examination is
a frequent inhabitant of shrubberies, but not a wild
plant; it is called " the tall" (H. elatum); if you
have it not at hand any other will do as well.

Its leaves have netted veins, are of an oval figure,
are placed in opposite pairs round the stem, upon
which they are seated without any stalks.   If you
rub them they emit a strong penetrating disagreeable
odour ;   the cause of which you may discover by
holding them against the light.   They will be seen to
be filled with transparent dots, crowded together and
so minute that you perhaps may require a magnifying
glass to discover them.   It does not always happen
that these dots are very small ;   on the contrary in
one of our wild species they are so large that the
leaves look as if bored full of holes, on which ac-

count the species has acquired the name of " the
perforated."

The flowers grow in loose clusters at the tops of
the shoots ; their calyx consists of five sepals which
are unequal in size, and which overlap each other
curiously at the edges when the flowers are very
young. To see this arrangement, cut a young flower-
bud across, and you will find that the two largest
sepals are on the outside of all (Plate V. 2. *fig.* 1**) ;
next one of these is a smaller, of which one edge is
covered by one of the large sepals, and the other lies
upon the edge of a still smaller one within it ; the
last is matched by a fifth of the same size as itself,
standing on the opposite side of the flower.

The petals are five, of a bright yellow, and very
large for the size of the flower. At the base of the
petals, and from below the pistil, arises a great number
of stamens of unequal lengths, with very fine yellow
filaments, and small roundish anthers ; if you take
hold of a few of these stamens with your fingers and
pull them, a cluster will separate from the rest (*fig.*
3.); and if you will pull the remainder they also will
come away in four other parcels ; so that the stamens
are really united into five different parcels, although
till you began to separate them you did not discover
it to be the fact. This is a curious circumstance,
to which you will find few parallels in other plants.

The pistil is an oblong body (*fig.* 4.) terminated by
three styles, each of which is tipped with a little
stigma. The inside of the ovary contains three cells,
in each of which is a multitude of ovules ; to speak

very correctly this pistil is composed of three carpels adhering to each other.

The ripe fruit is just like the pistil, except that it is darker-coloured, larger, and destitute of styles, which drop off shortly after the ovules are fertilized; it finally *(fig.* 7.), divides into three pieces or valves, each of which is one carpel; so that the adhesion between the carpels which took place when the flower was exceedingly young, does not cease till the fruit arrives at a state of dissolution. The seeds are very minute, but worth examining for their exceeding beauty. They are of an oval form, and up one side runs a curious crest *(fig.* 8.), which gives the seed something the appearance of an ancient helmet.

This plant represents the characters of an order called the *Tutsan tribe* (Hypericineæ), into which enter few other genera besides that which comprehends our wild flowers. Among them, however, are some, found in the tropical parts of America, called Vismias, which yield a resinous substance resembling gamboge. In fact, something of the same kind may be traced in many of the Tutsans themselves. True, gamboge is itself the produce of a tree of the natural order Guttiferæ, to which belongs, among others, the Mangosteen, which bears the most delicious fruit in the world; that natural order has an exceedingly strong affinity with the Tutsan tribe.

I need not recapitulate the characters of the orders explained in this letter, as they are so very distinct that you may safely be left on this occasion to your own ingenuity. In my next letter I shall present you with some very interesting subjects.

EXPLANATION OF PLATE V.

I. THE PASSION-FLOWER TRIBE.—1. A full-blown flower of the *Laurel-leaved Passion-flower; a* the involucre, *b* the sepals, *c* the petals.—2. A column of stamens ; *a.* the last ring of the rays.—3. A ripe fruit of the *Red-stemmed Passion-flower.*—4. The same cut through, to shew the manner in which the seeds are attached.—5. A seed, natural size, with its arillus on.—6. The same, with the arillus stripped back.—7. A seed magnified, with the arillus on.—8. The same with the arillus off; shewing the sculptured seed-coat.—9. A seed cut through ; the embryo is seen lying in the midst of albumen in small quantity.—10. An embryo extracted from the seed, with its broad leaf-like cotyledons, and small tapering radicle. (N.B. These are copied from drawings by Mr. Ferdinand Bauer.)

II. THE TUTSAN TRIBE.—1*. A flower-bud of *Tall Tutsan*, shewing the calyx.—1**. A view of the manner in which the sepals are respectively arranged when the bud is young.—2. A full-blown flower.—3. One of the five parcels of stamens.—4. A pistil.—5. A fruit three-quarters ripe.—6. The same cut through, to shew how the inside is arranged.—7. The ripe fruit separating into its constituent carpels ; which leave behind three pieces of their edge, in the shape of three narrow plates, to which the seeds once grew.—8. A seed very highly magnified.—9. A section of the same, shewing the two cotyledons, and *a,* the thickness of the crest.

LETTER VI.

THE MALLOW TRIBE—THE ORANGE TRIBE.

(Plate VI.)

WELL do I remember the pleasure I used to have, when a little fellow just sent to school, in gathering cheeses out of the hedges : it was my first step in Botany ; and it was not without pride that I found myself able to shew my less learned companions how to distinguish the plants that bore those delicacies. Many years after, when the cares and pleasures of life had blotted out all remembrance of the joys of childhood, I was passing a few days in Normandy, with my friend M. de P., when, one day, his little girls came running to me with their hands filled with fine plump *fromageons ;* I know not whether it was the association of ideas that the well-remembered word conjured up, or the sweet countenances of those dear children, joy painted in their black and sparkling eyes, and health in their rosy cheeks— but I ate their fromageons with as much delight as ever, and fancied them as superior to all the fruits of the garden in flavour, as they are in perfect symmetry of form. Only compare a vegetable cheese, with all that is exquisite in marking, or beautiful in arrangement in the works of man, and how poor

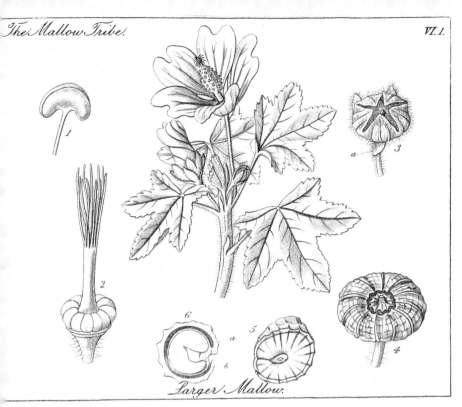

The Mallow Tribe.

VI. 1.

Larger Mallow.

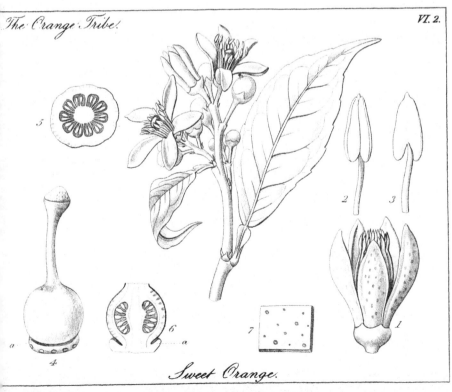

The Orange Tribe.

VI. 2.

Sweet Orange.

and contemptible do the latter appear! Not only,
when seeing it with the naked eye, are we struck
with admiration at the wondrous perfection and
skill with which so obscure a point in the creation
is constructed; but, using our microscope, surprise is converted into amazement when we behold
fresh beauties constantly revealed, as the magnifying power is increased, till at last, when the
latter reaches its limit, we find ourselves still re-
garding a lovely prospect, the horizon of which re-
cedes as we advance. Nor is it alone externally
that this inimitable beauty is to be discovered; cut
the cheese across, and every slice brings to view
cells, and partitions, and seeds, and embryos, arranged with an unvarying regularity, which would be
past belief if we did not know, from experience, how
far beyond all that the mind can conceive, is the
symmetry with which the works of Nature are constructed.

> Look, then, who list your gazeful eyes to feede
> With sighte of that is faire, look on the frame
> Of this wide universe, and therein reade
> The endless formes of creatures, which by name
> Thou canst not count; much less their natures aime;
> All which are made with wondrous wise intent,
> And all with admirable beauty blent.

I perceive, I have been talking of these curious
productions, as if you were already acquainted with
them; while it is quite possible that you are not. In
that case, step to the road-side, or to the first patch
of weeds you can meet with, and there you will be sure

to find what is called a Mallow ; we have two very
common sorts; one of which has small pink flowers;
the other, large striped purple ones; the latter is one
of our handsomest wild flowers, and is called "the
common," or, *the larger Mallow* (Malva sylvestris);
it is that you are to take as the subject of your
study.

This plant (Plate VI. 1.) grows two, or even three
feet high, in places where it is not cropped by cattle.
It has an erect branching stem, of a very pale
green, covered all over with longish hairs, which frequently spring from the surface of the stem in starry
(or, as we pedantically say, *stellate*) clusters.   The
leaves are roundish, and divided into about five shallow lobes, the border of which is notched; their veins
are netted. At the base of the leaf-stalk grows a pair
of small stipules, resembling scales.

From the bosom of the leaves spring the flowers
singly.   Below the calyx are placed three small
bracts, forming an involucre (*fig.* 3. *a.*).   The calyx
is composed of five sepals, joined together about
halfway; it is quite soft, with long delicate hairs.
Five large rosy-purple striped petals, each of which
has almost the figure of a wedge, and is notched at
the end, constitute the corolla, which spreads wide
open, when its proper time for unfolding arrives;
before that time, its petals were curiously twisted
together.

The stamens are very different from any we have
yet examined; they consist of a hollow column,
bearing, at its upper end, a great number of anthers,

each of which has a short filament (*fig.* 1.), and is of
a kidney shape, containing only one cell instead of
two, as is usual. Formerly Botanists were contented
to call this column, a column, and to inquire no
further; as if they thought it was some new and
special organ found only in such plants as the Mallow. At the present day, we are too curious to be
thus easily contented, and we must have the exact
nature of every part explained. This column, then,
is caused by the filaments growing fast together,
when they are very young, without being able to separate afterwards, except just at the top, where they
look like filaments. Suppose the stamens of the Tutsan were joined together in this way, when young,
you would have exactly such a column as is constantly produced in the Mallow.

The next object of examination is the pistil (*fig.*
2.); it is formed of several carpels, which grow together in a circle round a common centre, and so
form a sort of flat plate, from the middle of which
the styles arise. Like the filaments, the styles also
grow together at the bottom into a column, but they
soon separate again, and then you may tell, by
counting them, that each carpel has its own style,
for there is exactly as many styles as carpels.

Last comes our acquaintance the cheese, in the
shape of the nearly ripe fruit; we will suppose it to
be quite ripe (*fig.* 4.), for the sake of avoiding repeti-
tion. It consists of a number of dry carpels which
will separate readily from each other and from the
central body to which they were originally joined.

Each carpel (*fig.* 5.) contains one seed, with an embryo curiously doubled up and filling the whole of the cavity; hence as the carpels are all of the same size, and arranged with the most exact regularity on the same level, if a fruit is cut through it will present a singularly beautiful arrangement of the parts, which look like a vegetable star. In the centre, if the fruit is not ripe, is a solid circle from which eleven rays branch off at regular distances, each being sub-divided into two. Between the rays lie eleven embryos, the various convolutions of which, as cut through by the knife exhibit eleven areas of strange patterns. The kaleidoscope itself can produce nothing prettier than this, except in colour.

This account is that of all the Mallow tribe in most respects; and is quite sufficient to enable you to identify it: a power that it is useful to possess because the species are all perfectly innocent. The columnar stamens themselves suffice in a majority of cases.

The species yield a large quantity of a ropy transparent almost tasteless fluid, in all the green parts; and it is for this that the cheeses are sought with so much avidity. In India a daily use is made of the cheeses of a Malvaceous plant called *Ochro*, or in some places *Gobbo* (Hibiscus esculentus), the mucilage of which is in great request for thickening soups: it is even imported in a dry state into England. Marsh Mallow, which possesses this mucilaginous quality in a high degree, in all its parts, is a favourite material with many physicians, especially the French, for poultices.

We are, however, far from exhausting the properties

of the Mallow tribe, in talking of its mucilage. It
has many other and far more important good qualities.
In beauty it yields to no part of the vegetable world;
many of the plants called Hibiscus, are trees or shrubs
or herbs bearing flowers of the largest size, the most
exquisite proportions and the most striking colours ;
the common *Bladder ketmia* for example. One of
them, conspicuous for the brilliant crimson of its
flowers (Hibiscus Rosa Sinensis), is a vegetable shoe-
black ; the petals communicate a black stain to any
thing they touch, and the Chinese actually black
their shoes with them. Many varieties of it are
common in our hot-houses, so that you may try the
experiment on your own shoes if you like.

Cordage is also produced by several species, from
the tough vegetable fibres of their stem. But it is
the hairy clothing of the seeds of different plants
belonging to a genus that Botanists call Gossypium,
which is of such pre-eminent importance as to claim
for the Mallow tribe a rank in the vegetable kingdom
second only to Corn. That hairy substance is Cotton,
which for no conceivable purpose except to yield man
the means of clothing himself, is formed in prodigious abundance upon the back of the seeds of the
Cotton plants, whence it is torn by machinery and
afterwards cleaned and spun into thread. Faithless
travellers and credulous readers for a long time
caused the existence of the Barometz, or Scythian
Lamb, to be believed in, a creature said to be half
animal half plant ; the Cotton plant has far stronger
claims to the name of a vegetable sheep.

The simplicity of its characters and the remarkable arrangement of its stamens, render it unnecessary for me to dwell longer upon this natural order.

The next to which I shall direct your attention, is that of the Orange, in which I hope to make your little pupils take as much interest botanically, as they already do in a more practical way.

Oranges, Shaddocks, Limes, Lemons, Forbidden Fruit, and the like, are all produced by plants which represent a tribe perfectly distinct from the rest of the vegetable kingdom, and which we call the *Orange tribe.* They are all natives of countries warmer than this, and principally of the temperate parts of India; their fruit is in all cases eatable, although not always worth eating; their leaves and flowers all fragrant, and they are universally evergreens of beautiful appearance. The cells filled with oil, which you find in cutting the rind of an Orange, are met with in both leaves and flowers, to which they often give the appearance of being covered with little blisters; and, as usual, it is in them that the sweet odours are stored up.

To understand the structure of this interesting tribe, let us take the common sweet Orange, a plant or two of which is kept in every green-house, for the sake of the delicious fragrance of the flowers. It has leaves with netted veins, and filled with transparent spots (*fig.* 6.); they are always jointed just above the footstalk, so that each leaf will readily snap asunder into two pieces. The calyx is a little cup with five shallow teeth: so complete is the combina-

tion between the sepals of which it is composed.
Five white fleshy green-dotted petals are placed
within the calyx, and within them are ten or twelve
stamens that arise from under the pistil; these grow
together, in a way that explains well enough the real
nature of the column of the Mallow tribe, but that is
not characteristic of the Orange tribe itself.

The pistil (*fig.* 4.) is a round dark-green part, terminated by a thick style and stigma; around its
base is stretched (*fig.* 6, *a.*) a ring, out of the outside
of which the stamens originate. If you cut into the
ovary, you will find it contains several cells, in each
of which is a double row of ovules (*figs.* 5 and 6.).

Thus far there is nothing but a peculiar combination of parts, with all which you are already quite
familiar. But as soon as the ovary begins to grow
into a fruit, a great change comes over it: numbers
of the ovules perish; the thickest part of the rind
begins to separate from the lining, and finally becomes so loose that, as you know, it is easily peeled off;
and at the same time a great quantity of little pulpy
bags jut forwards into the cavity of each cell, becoming more and more watery, more and more acid,
and then more and more sweet, till at last the whole
substance of the fruit is a mass of sweet and delicious
pulp. The nature of these bags cannot be readily
seen in the Oranges of the shops, but if you examine
a bad Orange, such as is usually produced in a
green-house, the structure becomes most obvious.

Do not, however, suppose that the presence of
pulpy bags in the inside of the cells of the fruit is an

essential character of the Orange *tribe;* on the contrary, it is peculiar to the Orange *genus* (Citrus)
only.   In all the other genera the fruit is fleshy and
fragrant, but not pulpy in the inside.

So few plants related to the Orange are ever likely
to be met with by you, that I should only fatigue you
unprofitably in repeating their names.   Let us con-
clude, then, for this time with the essential character
of the Orange tribe, which, if reduced to its simplest
form, may be expressed thus :—

Leaves, flowers, and fruit, filled with transparent
receptacles of fragrant volatile oil.   Leaves jointed,
once at least, above the footstalk.   Stamens few and
hypogynous.   Fruit fleshy.

To compensate for the length of some of my former
letters this is unusually short; for which I suspect
you will be inclined to thank rather than to blame
me.

EXPLANATION OF PLATE VI.

I. The Mallow Tribe.—A sprig of *the larger Mallow* (Malva
sylvestris).—1. An anther with the upper part of the filament.—2. A
pistil, shewing the ovary, column-like base of the styles, and the styles
themselves.—3. A calyx closed over the growing fruit, with leaves of
the involucre at *a.*—4. A ripe fruit ready to separate into distinct carpels.—5. One of the carpels separated from the rest, and seen from its
side.—6. The same, split in halves, to shew the embryo; from the
position in which the embryo lies, only one of the cotyledons can be seen
at *a; b* is the radicle.

II. The Orange Tribe.—A twig of the common *sweet Orange.*
—1. A flower, shewing the calyx, the petals spotted with oil, and the

tips of the stamens.—2. A stamen, with the anther viewed in front.—
3. The same from behind.—4. A pistil, with its disk at a ; at the lower
edge of the ring or disk are seen the scars caused by the separation of
the stamens.—5. A cross section of the ovary, shewing the numerous
cells and double lines of ovules.—6. A perpendicular section of the
same ; even at this young period you may see in the rind of the ovary
the receptacles in which oil is beginning to secrete ; a the disk.—7. A
portion of a leaf magnified so as to shew the oil cells.

LETTER VII.

(Plate VII.)

FROM the earliest period of your familiarity with a
garden, you must have been acquainted with those
sweet aromatic flowers called *Pinks*, *Piccotees*, and
Carnations, and you must have admired their beautiful stripes, and the symmetry with which their
petals are arranged. It is also not improbable that
you have some knowledge of a mean weed, called
Chickweed (Stellaria media), which inhabits every
neglected corner of your garden; *Corn Cockle*
(Agrostemma Githago), *Bachelors Buttons* (Lychnis
dioica), *Ragged Robin* (Lychnis flos Cuculi), and
many species of *Catchfly* (Silene) are also pretty
flowers, that you will easily procure either by hunt-
ing for them in the fields, or by inquiry after them
in gardens.

All these agree with each other in a number of characters which are so remarkable as to divide them
from all other plants, and to cause them to be established as a distinct natural order called the *Chick-*

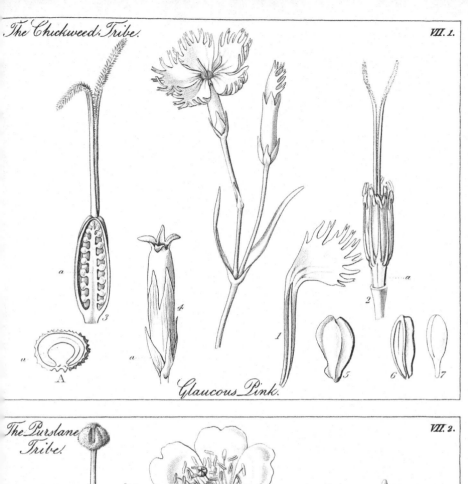

Glaucous Pink.

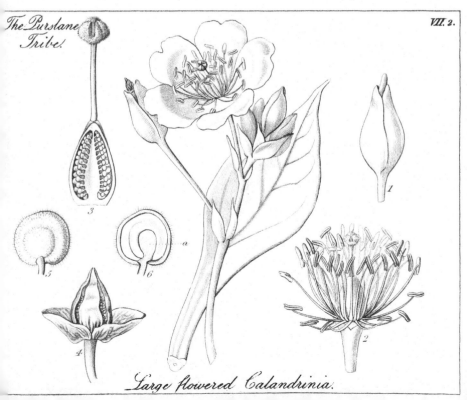

Large flowered Calandrinia.

weed tribe; which is composed for the chief part of
plants of little interest or beauty, among which there
is not a single species with unwholesome properties.
Uninteresting as many of them are, they are so common that every one who pretends to botanical knowledge must learn how to recognize them, even if it
were not for the sake of the few kinds, that, like the
Pink, are our familiar acquaintances.

To understand the structure of the Chickweed
tribe, I shall not ask you to take the Chickweed
itself, because it is a plant with very small flowers;
let us rather seek some species in which all the parts
can be easily seen; as a Pink for instance. Here is a
pretty species, the *glaucous Pink* (Dianthus glaucus) of
Scotland; if you have it not in your garden, any
other will do as well, provided it is not double.

This little herb is called glaucous, from a Latin
word signifying blueish-grey, because its leaves, like
those of many other pinks, have such a colour in a
remarkable degree. Its stems are very much swoln
at the joints where the leaves are set on. The leaves
are exceedingly narrow, undivided, and rather rough
at their edge; they have only one single vein which
runs through them from one end to the other. How
then are we to ascertain whether this plant is Exogenous or not? for there is nothing here to shew
whether the veins have a netted structure; there is
apparently only one vein to examine. I must confess this looks very like a difficulty; and I dare say
you will now suppose that the time has come when
you must have recourse to patience and a microscope

H

to learn whether there are two cotyledons in the embryo or only one (see page 13); believe me, however, we have not yet arrived at so disheartening a point. There are in fact many ways of shewing you how to determine whether this is an Exogenous plant or not, without counting the seed-leaves. That which I select is one of the easiest to understand; but I must first mention a few matters that I have not hitherto touched upon.

You are no doubt acquainted with some of the idle tales that are told by the ancient poets, of people being changed into trees, or animals, or rocks; one young lady for example, not only cried her eyes out, but was altogether changed into a running stream, and another was transformed into a spider, because she dared to emulate the goddess of wisdom in tent-stitch; these occurrences they called Metamorphoses, a name which Botanists have borrowed for something of a similar nature which really does happen in plants. Hitherto I have always spoken of the different parts of the flower as so many totally distinct organs, and it is certainly true that the petals, stamens, and pistil have very different offices to perform. But, at the same time, it seems quite certain that all those, and several other parts, are in a very great degree constructed like leaves; that at a very early period, when they were first formed, they were absolutely the same as leaves of the same age; that it is only after they have been growing for some time that they begin to assume the characters under which they finally appear; and that consequently they are very often found resuming the appearance of common leaves if any thing occurs to interfere with their intended structure before it is entirely fixed. Thus we find leaves in the place of petals, or as they say petals metamorphosed into leaves, in some kinds of double Tulips ; sepals and pistils often changed to leaves in double Roses ; all the parts of the flower turned into leaves in other plants ; and a multitude of similar cases with which the Botanist is acquainted. You will find all this so fully demonstrated in my Introduction to Botany, p. 504, or in other works of a like nature, that there is no doubt of the fact ; and the doctrine is now the foundation of modern views of the real structure of flowers and fruit. It is not my plan to enter into such questions in these letters, but I could not avoid calling your attention to the circumstance.

Now mark the practical application of this knowledge. If the parts of the flower are only leaves in a particular state, any of those parts in which veins can be discovered will serve to shew the arrangement of the veins as well as the true leaves themselves. In the Pink the petals are fully expanded, and full of veins ; they are therefore fitting objects to examine ; and their structure will tell us whether the Pink is Exogenous or not. You will find them distinctly netted, and thus that question is set at rest.

This then, which is an Exogenous plant, has opposite undivided leaves seated on the swoln joints of the stem. The calyx consists of a tube composed of five sepals joined together and separated only near the points. Five petals arise from within them, each of which has a stalk and a blade ; the stalks or claws as they are called *(ungues)* are very narrow and stand side by side within the calyx ; the blades are much expanded and irregularly lacerated at the end.

Stamens there are ten, rising from beneath the ovary, out of a short stalk (Plate VII. 1. *fig.* 2. *a.*). The ovary is superior, and contains but one cell, in the centre of which is a slender receptacle *(fig.* 3. *a.)*, covered with a great many ovules. The styles are two, each terminating insensibly in very narrow fringed stigmas.

The fruit becomes a dry case or capsule, opening at the point with four teeth or valves *(fig.* 4.). The structure of the seed is variable, and not important for our present purpose.

Such is the character of the Pink, and such to a great extent is that of the tribe it represents. It may be said to consist in these marks. *Stems swoln at the joints; leaves opposite and undivided; stamens few and hypogynous; ovary with many styles, one cell, and a central receptacle covered with ovules.* Nothing like this has been previously shewn to you. The genera are very uniform in their structure, and are distinguished by marks that every one may observe. Two divisions are formed, one of which has the sepals united into a tube, the other has them all distinct.

In the first division is found the *Pink*, which is

known by the bracts at the base of its calyx (*fig.* 4.
a.) ; and some others, of which the following are the
most remarkable ; Silene or *Catchfly* has three styles
and a little crest at the top of the stalk of each petal ;
it derives its English name from its often secreting a
viscid matter in which flies are caught. The *Cockle*
(Agrostemma) has five styles and undivided petals.
Lychnis, to which the *Ragged Robin* (L. flos Cuculi),
and *Batchelors Buttons* (L. dioica) belong, to say
nothing of the splendid *Chalcedonian Lychnis* of the
gardens, has five styles and divided petals.

To the second division we refer *Chickweed* (Stellaria),
which has three styles and two-lobed petals, *Sandwort*
(Arenaria), which has three styles and undivided
petals, and *Mouse-ear Chickweed* (Cerastium), which
has five styles and a curiously shaped taper seed-
case with ten teeth.

So little interest attaches to these plants that I leave
them here, and proceed to notice another order
which resembles them in many respects ; and which
is upon the whole more beautiful.

Imagine you have a Chickweed with *two sepals and
one style*, all the other points of structure in the flower
remaining the same ; and you will have an order
that, while it seems to resemble the Chickweed tribe,
so much as to be almost identical with it, nevertheless differs in several important particulars in the
manner of growth. Not only are the leaves alternate instead of opposite, and the stem destitute of
swellings at the joints, but there is a constant disposition on the part of the stems and leaves to become

fleshy, or as it is usually called succulent. Plants
thus constructed belong to the *Purslane tribe,* so called
because the now-forgotten Purslane, which was once
in much esteem as a salad, belong to it. They are
often remarkably beautiful on account of their bright
red or yellow flowers; are always harmless and
wholesome in their properties; but are sometimes,
when the petals are small, mean-looking herbs.

You cannot take a better specimen of the Purslane
tribe than the *large-flowered Calandrinia* (Plate VII.
2.), a Chilian plant that is now a good deal cultivated. Its soft and succulent stems and leaves are
exceedingly dissimilar to those parts in the Chickweed tribe. The calyx is formed of two leaves (*figs.* 1.
and 4.); there are five petals which pay their homage
to the sun by unfolding their crimson drapery
beneath his earliest beams, and rolling up again as
soon as the light of his countenance is withdrawn.
A considerable number of stamens succeed the
petals; and in the centre is an ovary with one cell
and a central receptacle covered with ovules; it is
surmounted with a style which ends in a broad hairy
lobed stigma. The fruit when ripe splits into four
valves (*fig.* 4.); which allow the escape of a number
of black seeds (*figs.* 5. and 6.), altogether resembling in their internal structure the seeds of most of
the Chickweed tribe.

The obscure little *Water Chickweed* (Montia fon-
tana); a neat looking American genus of hardy plants
called Claytonia; the *Purslanes* themselves (Portulaca),
many of which are very handsome; and the Chilian

Calandrinia are the most common representatives of the order.

While the principal part of the plants belonging to the Chickweed tribe is found in meadows, or shady spots, or in situations and climates where they are abundantly supplied with moisture, the Purslane tribe, on the contrary, chiefly rejoices in hot dry exposed places, where they will flourish at a time when every thing else has fallen a victim to drought and heat. They owe this power to the peculiar nature of their stems and leaves; which, as I have already mentioned, are succulent. They require, in consequence, a peculiar mode of cultivation, which I will now explain to you.

When I gave you in my first letter a brief account of the minute and beautiful arrangements by which leaves are able to perform their vital actions, I omitted to say anything about their breathing-pores or STOMATA. The time has now come when I must tell you what they are.

All leaves are covered by an exceedingly thin and delicate skin, which you may often peel off. If you put a small piece of this in water, and look at it with a microscope against the light, you will remark a number of very small roundish or oval spaces, through the middle of which a line passes. And if you have patience enough to look at a great many of them, you will in time perceive some of them opening a sort of mouth at the place where the line is. These are what are called breathing-pores; they are organs by which the leaves inhale and exhale air, or vapour

suspended in air. In some leaves they are extremely
large and numerous, and such leaves perspire through
the breathing-pores in very great quantity, as the
Vine ; in others they are so small or so few, as to admit of scarcely any perspiration, as in the Purslane,
and in succulent plants in general. It is thought,
indeed, that the character of succulence is owing
to leaves being unable to get rid of the water ab-
sorbed by the roots, and so becoming, as it were,
dropsical.

Attention to this difference in the power of perspiring in different plants is one of the keys to a
knowledge of the right method of cultivating them,
and it has been applied for years to succulent plants,
by keeping them in what is called a dry stove; that is
to say in a hot-house, the air of which is kept dry by
refraining from watering the floors or earthen pots in
winter. Succulent plants are, when in a state of
rapid growth, so imperfectly supplied with the means
of perspiring, that they require all the assistance
which can be obtained from a dry atmosphere, to be
able to part, by the leaves, with the water that is impelled into them through their roots ; and consequently, if ever, in a rapidly growing state, they are
kept in a damp atmosphere, they become dropsical
and unhealthy, or soon decay. But in the winter,
the little power of perspiration which they possessed
in the full vigour of their summer growth is very
much diminished, and is in fact reduced to almost
nothing. Their roots will nevertheless go on absorb-
ing moisture from the soil as long as the soil contains

any, and this, if the moisture is in much quantity, is certain to produce decay and death, because the excess of water, which cannot be afterwards parted with by the leaves, becomes putrid. Such being the case, it is found necessary to deprive succulent plants of all water at their roots in winter, and to leave them for support to the vapour which always will exist in the air at that season in a climate like that of England.

It is for these reasons that succulent plants succeed so much better than others in sitting-rooms. In such situations, plants are killed by want of light, and want of moisture in the air; for the air of all sitting-rooms is necessarily very dry. But succulent plants apparently require less light than most other plants, and are certainly benefited rather than injured by a dry atmosphere. I should therefore advise you, if you are anxious to have a garden in your sitting apartment, to fill it with succulent plants, to the exclusion of all others.

The Purslane tribe is far from being the only one in which succulent species are found; they might exist in any, and in fact do in a great many natural orders, the majority of whose species are not succulent; some orders, however, abound in them more than others, as for example, the *Houseleeks* (Semper-viveæ), the *Torch-thistles* (Cacteæ), the *Asclepiadeæ*, the *Euphorbias* (Euphorbiaceæ), and the *Asphodels* (Asphodeleæ).

But it is time that an end were put to this letter, especially as the next must, I fear, be a very long one.

106

EXPLANATION OF PLATE VII.

I. The Chickweed Tribe.—A piece of the *Glaucous Pink.*—1. A petal apart; on its stalk is a curious double raised plate.—2. The stamens and two stigmas, as seen when the calyx and corolla are cut off; *a* is the stalk of the ovary.—3. A perpendicular section of an ovary, shewing the central receptacle, *a*, on which the ovules grow.—4. The ripe fruit invested with its calyx, and having the bracts *a* at the base.—5. A seed.—6. The same divided perpendicularly, so as to shew the embryo.—7. An embryo taken out of the seed.—A. a seed of another plant, *Batchelors Buttons* (Lychnis dioica), cut through perpendicularly, and shewing the curved embryo; *a* albumen.

II. The Purslane Tribe.—A leaf and a portion of a flower-branch of *Large-flowered Calandrinia.*—1. A flower-bud, with the sepals, and the corolla peeping through it.—2. The stamens and pistil.—3. A perpendicular section of the ovary, shewing the central receptacle on which the ovules grow.—4. A fruit beginning to open, with the two sepals by which it is protected.—5. A seed, with a portion of its stalk. —6. The same cut through, to shew the embryo and the albumen, *a.*

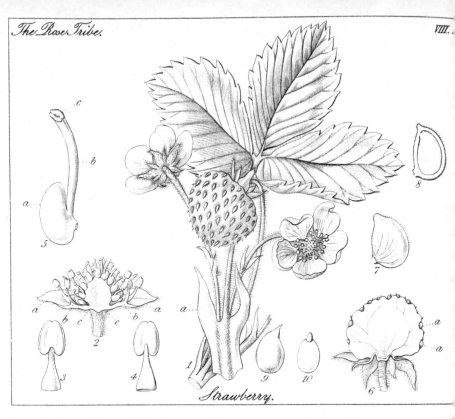

Strawberry.

Narrow-leaved Restharrow.

LETTER VIII.

THE ROSE TRIBE—BUDDING AND GRAFTING— THE PEA TRIBE.

(Plate VIII.)

You, perhaps, remember, when you were beginning to study Botany, how you fell into the error of supposing that the Strawberry belongs to the Crowfoot tribe, and that I then explained to you the cause of your mistake. In this letter I propose to take the Strawberry, for the purpose of illustrating the natural order to which it belongs, which we call the *Rose tribe*, because the Rose is one of the genera comprehended in it. I am sure, that the mere mention of this favourite plant, will ensure your attention to the history of its relations.

The *Strawberry* (Fragaria) is a herb with three-parted leaves, and a pair of large membranous stipules at their base (Plate VIII. 1. 1. *a*.). The veins of the leaves are netted. When the Strawberry plant is about to multiply itself, it puts forth naked shoots of two sorts : one of them is prostrate on the ground, and ends in a tuft of leaves, which root into the soil, thus forming a new plant; or, as it is technically called, a *runner;* the other kind of shoot grows nearly erect, and bears, at its end, a tuft of flowers, which afterwards become fruit; or at least what is commonly called so.

The calyx of the Strawberry is a flat green hairy part, having ten divisions; it is, therefore, caused by the union of ten sepals, five of which are on the outside of the others, and smaller. As you become more and more acquainted with Botany, you will find that it is an extremely common thing for the parts of the flower to consist each of several rows. This is the first time I have mentioned it, and it is the first time it has occurred in the calyx; but you have already seen it in the rays of the Passion-flower, in both the stamens and carpels of Crowfoots, and in the stamens of many of the other orders we have passed by.

The corolla consists of five petals.

The stamens are very numerous, and are placed in a crowded ring round the pistil, as in the Crowfoot; but, you will observe, that they grow out of the side of the calyx (*fig.* 2.), and not from beneath the carpels.

The pistil of a Strawberry is very much like that of a Crowfoot: it consists of a number of carpels, arranged in many rows, and with great order, upon a central receptacle; each carpel has a style, which arises from below its point (*fig.* 5.), and terminates in a slightly-lobed stigma. In the inside of the ovary is one single ovule. With the flower, the resemblance between the Crowfoot and the Strawberry ceases.

You will almost wonder, now that you know how
the young flower of a Strawberry is constructed, how
so singular a fruit is to be formed out of such materials; especially if you should have chanced to meet
with the ingenious explanation given of it by some Botanist, whose name I forget, that it is a *berry with its
seeds on its outside.* Many and strange are often the
changes that take place in the organization of a pistil
in the course of its transformation into a fruit : and
they are highly curious in this case. If you would
really understand them, you should watch the Strawberry in the progress of its growth. You would then
see that the first occurrence after the petals have fallen off, and the calyx closed on the tender fruit, consists in the receptacle of the carpels beginning to
swell; and, shortly after, in the carpels themselves
gaining a greater size and a more shining appearance ; while, at the same time, their styles begin to
shrivel up. At a more advanced stage, the carpels
are found but little augmented in size, while the receptacle has increased so very much in dimensions,
that the carpels are beginning to be separated by it,
and the surface of the receptacle can be distinctly
seen between them. A little older, and the carpels
seem scattered, in an irregular manner, over the surface of the receptacle, which has become soft and
juicy, while they have remained almost stationary in
size. All along the swelling receptacle has been
pushing the calyx aside, as being no longer of use
to it; and, at last, you scarcely remark the calyx,
in consequence of the much greater size of the re-

ceptacle. This part finally gains a crimson colour,
swells more and more rapidly, acquires sweetness
and softness, and at last is the delicious fruit you
are so well acquainted with :—in that, its final state,
the carpels are scattered over its surface in the form
of minute grains, looking like seeds, for which they
are usually mistaken. You, however, know better
than to fall into this common error ; for you have
seen, that, at first, they had each a style and stigma,
which seeds never have; and you can now, by cutting them open (*fig.* 8.) detect the seed (*fig.* 9.) lying
in the inside of the shell of the carpel. The Strawberry is, therefore, not exactly a fruit ; but is merely
a fleshy receptacle bearing fruit ; the true fruit being
the ripe carpels.

Cinquefoils (Potentilla), are little herbs, usually
with pretty yellow flowers, found growing on banks
and on commons, among the short grass ; one of them
is called *Silver-weed* (P. anscrina), on account of the
white, and almost metallic, appearance of the under-
side of its leaves. They are so like the Strawberry
in flower, that there is no Botanist who could tell you
how to distinguish them in that state. But they do
not bear Strawberries ; that is to say, after their
flower is withered the receptacle does not become soft
and pulpy ; but it always remains hard and dry, and
is completely hidden by the carpels. This plant,
then, must be very nearly related to the Strawberry.

The *Raspberry* and *Bramble* (Rubus), also claim
kindred with the Strawberry, because of their like-
ness to it. They are shrubby plants, having their

stems covered with hard hooked prickles ; in this respect they differ from the Strawberry. Their leaves
are divided, in various ways, according to their
kinds, and have large stipules at the bottom of their
stalks ; here they agree with it. Their calyx has
only five divisions instead of ten ; which is a difference ; but their petals are five ; the stamens numerous, and arising out of the side of the calyx ; and
their pistil composed of a number of carpels arising
out of a central receptacle ; these, again, are resemblances with the Strawberry in important points.
Let us examine the fruit. The Raspberry has a dry
core, off which you may pull the little thimble-like
fruit, and you will not find any of the dry grains
which stick upon the outside of the Strawberry. But,
look again ; what are the little dry threads that you
see rising from the centre of a multitude of little projections with which the whole surface of the Raspberry
is covered ? Surely they are styles ; and, if so, the
projections out of which they grow, must be carpels
in a ripe state. This is really the case ; the carpels
of the Raspberry, instead of remaining dry as they
become ripe, swell, and acquire a soft pulpy coat,
which, in time, becomes red : they are crowded so
closely, that, by degrees, they press upon each other,
and at last, all grow together into the thimble-shaped
part which you eat ; in order to gain this succulent
state, they are forced to rob the receptacle of all its
juice, and, in the end, separate from it, so that when
you gather the Raspberry, you throw away the receptacle under the name of core, never suspecting

that it is the very part you had just before been feast-
ing upon in the Strawberry. In the one case, the
receptacle robs the carpels of all their juice, in order
to become gorged and bloated at their expense; in
the other case, the carpels act in the same selfish
manner upon the receptacle.

Another plant very like Cinquefoil is *Avens* (Geum),
one species of which, called the *Herb Bennet* (G.
urbanum), is common enough in hedge rows. It
grows a foot or two high, and has the leaves upon
the stem in three lobes, while those at the bottom of
it are in many divisions. The flower of this plant is
so extremely like that of Cinquefoil, that you could
not distinguish the two. But its fruit is a sort of
bur, composed of innumerable stiff bristles, which
all spring from one common centre, and are terminated by hooks at the point. The bristles are the
styles which, like those of the Geranium, have grown
stiff and long; and the hooks are the hardened points
of them where they have curved back, and separated
from an upper portion which drops off. The central
part is a mass of carpels, the receptacle of which is
hard and dry.

A beautiful mountain plant, frequently met with in
the Alps of Europe, called *Mountain Avens* (Sieversia
montana), and often seen in gardens, where it is cultivated for its large yellow flowers, and its diminutive stems covered with large deeply-lobed leaves,
offers a further instance of the changes which the
fruit of the Rose tribe undergoes between youth and
old age. When the fruit of Mountain Avens is ripe,

it looks like a silken plume springing out of the cup
of the calyx, and as it waves about in the wind one
may almost fancy it a tuft of feathers accidentally
fastened to the flower-stalk. A botanical examination dispels the illusion, and shews that the appearance is caused by the carpels having preserved their
styles, which become very long and are covered all
over with loose silky hair, which has grown since
they were young. A similar phænomenon occurs in
the *Virgin's bower* (Clematis), and in the *Pasque flower*
(Anemone); but the most remarkable instance of the
production of hairs so as to change the whole appearance of a part, is met with in the *Venetian Sumach*
(Rhus Cotinus), which the French call *Arbre à per-
ruque* or the Wig Tree. You have perhaps seen this
plant, which is by no means uncommon in shrubberies, "shaking its hoary locks" at you as the breeze
waved the branches, and set the wigs in motion, in
the midst of a crowd of blood-stained leaves; if you
have not, I would advise you to seek for it in the
autumn, at which season only it wears its wig; in the
spring and summer it does not want it and will not put
it on. The explanation of its strange appearance I
cannot give better than in the words of Professor De
Candolle. "The panicle (that is the cluster of
flowers) of Rhus Cotinus is almost entirely smooth at
the flowering season; after that period all the flower-
stalks which bear fruit, continue to remain smooth or
scarcely downy; but, on the contrary, on those
whose fruit is not formed, and they constitute the
greatest number, there appears a great quantity of

scattered hairs, which give them a shaggy aspect, on
which account gardeners have named the plant the
Wig Tree; it is probable that this excessive production
of hairs is caused by the sap, which was destined to
nourish the fruit, not having any employment in those
stalks the fruit of which is not formed, and expending
itself in the production of this extraordinary quantity
of hair."

The genera hitherto mentioned are upon the
whole better adapted to give a student a correct
notion of the character of the Rose tribe, than the
Rose itself. It is now necessary that we should examine this charming flower, in the construction of
which you will find as much to admire as in its external attractions. The leaves and stems of Roses
are sufficiently like those of the Bramble, to render
it unnecessary for me to insist upon any peculiarities
in those parts. In the flower much seems to differ,
although in reality but little essential difference
exists between Roses and other Rosaceous plants. It
has a calyx of five divisions, some of which are very
like small leaves. To these succeed five petals; and
within the latter is a great number of stamens, which
grow from the side of the calyx. You will not at
first sight perceive any pistils : in the centre, indeed,
is a tuft of stigmas, but no ovaries are visible ; upon
further search, however, you may discover, especially if you press the flower forcibly between the
finger and thumb, that the styles project through the
neck of an oblong green body, which being below
and on the outside of the calyx looks like an inferior

ovary   If you split the flower perpendicularly, you
will then perceive that this body, that looks like an
inferior ovary, is in reality the tube of the calyx,
which is contracted at the place where the stamens
originate, into a narrow orifice, through which the
tops of the styles protrude; and that the ovaries are
included within that tube, forming as usual the bottom of the styles.   Perhaps you will gain a clearer
notion of this if you suppose that part of the tube of
the calyx of the Strawberry, which is included
between the letters *b* and *c* in the accompanying
figure (*fig.* 2.) to be lengthened very much, while
all the other parts retain their size and position; in
that case the carpels, with their receptacle, might
become much shorter than the tube of the calyx,
instead of being longer, and if the latter were contracted at its mouth, no part of the carpels would be
visible except the tops of the styles and the stigmas.
The ripe fruit, or hep of the Rose, is nothing more
than the same tube of the calyx turned red and
fleshy, the sepals, and petals, and stamens having
dropped off; in its inside will be found the carpels
changed to bony grains, covered with coarse stiff
hairs.

Thus are constructed those plants which most
exactly belong to the Rose tribe.   They are all
harmless, and when they are sufficiently agreeable to
the palate, eatable; very often, however, their juice
is so very austere and astringent that no use can be
made of them except in medicine.   They have sometimes been employed in domestic medicine, especially

Herb Bennet, the roots of which have been found by some physicians as valuable in the cure of fevers as the Jesuits' bark itself.

Very nearly related to the Rose tribe are two other sets of plants, which by some are reckoned mere subdivisions of it, though by others they have been considered distinct natural orders. We need not trouble ourselves with that inquiry; I dare say you will prefer to know what they are and how they are characterized.

The first of these is the *Apple tribe*, to which belong all those plants which agree with the Rose tribe in every thing but the carpels being distinct and superior; in lieu of which they have the carpels united and adhering to the tube of the calyx. Take an Apple tree in flower, as an example of this. It is a plant with leaves having netted veins, and stipules at their base. The calyx has five divisions, the petals are five, and there is a great many stamens growing out of the side of the calyx. In the centre you will find five styles; but their ovaries, instead of being merely enclosed within the tube of the calyx, adhere and form one body with it. It is this circumstance that gives rise to all the difference that you find in the fruit itself. An apple is a large fleshy body having at one end what is called an eye; which is in reality the remains of the calyx surrounding the withered stamens. The principal part of the flesh is the tube of the calyx, but the central part is the carpels, also grown fleshy, and at this period undistinguishable from the calyx itself; that their number

was five is shewn by the five cavities in the centre of
the fruit, each of which contains one or two seeds.
Now it is obvious, if this description be carefully considered, that the fruit is the only thing by which the
Apple is known from a Rosaceous plant. The same
kind of structure is found in the *Pear*, and the
Quince, and the *Mountain Ash*. In the *Medlar* and
the *Hawthorn* it seems as if the fruit contained two
or three stones instead of the open cavities of the
Apple; but in reality the only peculiarity in those
fruits consists in the lining of their cavities being
bony instead of thin and papery, as you may easily
satisfy yourself by looking at their flowers. All the
plants of this tribe are as harmless as the genuine
species of the Rose tribe itself.

The other group to which I have referred is the
Almond tribe. This is less different in structure
than the Apple tribe, but more dissimilar in sensible
properties. It consists of species which have all the
essential parts of structure of a common Rosaceous
plant, but which bear fruit like that of a Plum. The
Plum-tree itself, for example, has leaves with netted
veins and stipules at their base; a calyx of five
parts; five petals; and a great number of stamens
arising out of the sides of the calyx. But in room of
many carpels, there is only one, and that one changes
to a fleshy body, containing one single seed, enclosed
in a hard stone. The hard stone is the lining of the
cell of the carpel, separated from the fleshy rind
that is on its outside. This kind of fruit is called
a DRUPE. What is found in the Plum exists

equally, and but little modified, in the *Apricot, Peach, Nectarine, Almond,* and *Cherry,* all which are species of the Almond tribe. They would not perhaps be separated by Botanists into a distinct natural order, upon so slight a character, as the fruit being a Drupe, if that circumstance were not accompanied by a difference in the qualities of such plants. Instead of being perfectly wholesome, they are in many cases highly poisonous, as the *Common Laurel* (Prunus Laurocerasus), the leaves of which yield the dangerous infusion called laurel-water. This is owing to the presence of a volatile principle called Prussic acid, which in its concentrated state is one of the most dangerous of poisons. It is it which gives the well known flavour to Almonds, Ratafia, and the liqueurs called Maraschino, Kirchenwasser, and Mandel amara, and which is so often employed to mix with creams. Do not however suppose, that on this account there is any real danger in eating the fruit of Cherries, Plums, Peaches, and the like; in those fruits the prussic acid exists in such very minute quantity, as to be incapable of producing any deleterious effects. Nature has provided amply against the ill effects of such an insidious enemy, by rendering its presence instantly perceptible by an intensity of flavour that cannot be mistaken. There are those, indeed, who have condemned the whole tribe for the qualities of a few, and who have gone so far as to assert, that the dried leaves of the common Sloe were poisonous. It is not probable that the green leaves of that plant would produce any seriously bad effect; and it is certain that when dried they would lose what little poison they may have possessed when green, because the prussic acid principle is so volatile as to be immediately dispersed by the mere exposure of the leaves to heat.

Plants of the Almond tribe have also this peculiarity in which they differ from the Roses; their bark yields gum; as you may see upon the cracked branches of diseased Cherry and Peach trees. Like the Roses themselves, they have also a great deal of astringency; and their bark has been used in the cure of agues and fevers with considerable success.

If you analyze the characters of these three orders, you will find that their differences may be expressed thus :—

Ovary superior. { Many Carpels.—*The Rose Tribe.*
{ One Carpel.—*The Almond Tribe.*
Ovary inferior. . . . *The Apple Tribe.*

The woody species of these three natural orders are objects of such universal cultivation, Roses for their odour and beauty, Peaches, Apples, &c., for their utility as fruit-trees, that I cannot do better than explain to you the principles upon which the operations of budding and grafting, by which they are propagated, are conducted. If you do not care for multiplying Apples and Pears, I dare say you would at least be amused in making one kind of Rose grow upon another, and in converting the wild Briars of the neighbouring hedges into objects of greater beauty.

The gardener's operations of budding and grafting, depend for success upon the fact, that a portion of

one tree will grow upon another if skilfully applied to it. There are those who have believed
that a piece of one plant would grow upon *any*
other; and that Roses might be budded upon Black
Currant bushes or Pomegranates; this, however, is
untrue. It is a certain fact, that one tree will grow
upon another only when the two are *very closely*
allied in structure. Thus a Pear will grow upon a
Medlar or a Mountain Ash, but not so well as on some
other Pear; a Rose will grow upon any other Rose,
but not upon an Apple. This is a fundamental rule.

In the next place, the stems of all plants consist of
buds, and of the part that bears them; the latter has
no power of growing without the former, but the former can grow without the latter. For example, if I
plant a portion of a stem deprived of its buds, it will
die, notwithstanding all the care I can take to preserve it; but if I take the bud of a plant without
any stem, and place it in earth. it will grow, if due
precautions are employed; this shews, that the property of increasing a plant resides in the buds exclusively. It is not, however, *necessary* to separate
the bud from the stem; on the contrary, the two
taken together are frequently employed, when they
are called cuttings; if the bud alone is employed, it
retains its own name. These are the next points to
attend to.

Thirdly, both cuttings and buds will grow in other
media than earth, as for instance, in water, or damp
moss, or in any material which is capable of furnish-
ing them with a constant supply of moisture and

food. Nothing is more proper to furnish this than
the stem of a tree in which the sap is in motion ; because sap is moist, and at the same time, is the food
of plants ready prepared for them to consume it.

Hence, if a cutting, or a bud be carefully planted
in the stem of a tree, they will grow ; their roots will
be insinuated beneath the bark, and form new wood,
and so strong an adhesion will take place between
them, that no force can afterwards separate them.

Thus, if you would bud one plant upon another,
the plan is this ; it must be done when the sap is most
in motion, and when the bark can be easily divided from
the wood ; you then cut a narrow slice off the branch
which you wish to increase, taking care that it has a
well-formed bud upon it ; with a smart jerk, the small
quantity of wood that adheres to the slice may be taken
off, leaving nothing but the bud attached to the
bark.   Then, with a sharp knife, an incision must be
made lengthwise through the bark of the plant in which
your bud is to grow, and at the upper end of that incision a transverse cut, so that the two will together
form the letter T.   The bark is next to be lifted up
on each side of the longitudinal incision, and the
bud, with its adhering bark, is to be slipped in.   If
the whole is then bound up with worsted or matting,
a union will take place between the bud and the
branch into which it is inserted, and a new plant
will be created.

If you would graft one plant on another, you must
follow a different plan ; for grafting is effected with
cuttings, and cuttings cannot be so conveniently slipped

beneath bark, except at one end. There are many ways of performing the operation of grafting, one of which is the following :—You cut off the upper end of the branch on which you wish to insert a cutting; and you then pare the end of the branch flat, equally on both sides, till it resembles a long wedge. This done, you slit the lower end of the cutting, and pare away its wood to two flat faces, which correspond with the faces of the branch on which it is to be made to grow. You are then to fit the cutting accurately to the branch, taking care that the bark of both touches exactly, and if they are bound together with worsted or matting, as before, the operation is completed. As however by this method the cutting is less readily supplied with sap than in the case of buds inserted beneath the bark, it might perish before any union between itself and the branch could take place; to prevent this, it is usual to surround it with a thick coating of clay, which adheres to the surface of both branch and cutting, prevents evaporation, and also keeps the two more firmly applied to each other.

These are the simple means by which such important operations as the multiplication of rare and valuable plants, by making pieces of them grow upon those which are worthless, are performed. If we possessed no such power, it would be almost useless to occupy ourselves with improving the quality of fruit trees, for we should be unable to perpetuate them.

Here I would recommend you to pause for the present; and when what relates to the Rose Tribe is fully mastered, to proceed to the second half of this long letter.

There is, perhaps, no natural order which is more easily recognized than the *Pea tribe*, nor one in which greater interest is usually taken ; it is so rich in plants useful for food, as the *Pea*, *Bean*, &c. or for forage, as *Clover* and *Lucerne;* or dyes, as *Indigo* and *Logwood ;* or timber, as *Brazilwood*, *Rosewood*, and the *American Locust Trees ;* or medicine, as the *Senna* plant ; or gums, as the *Arabian Acacia ;* and attractive for their beauty, as *Robinias*, *Laburnums*, *Bladder Sennas*, and the noble tropical species of Butea, Jonesia, and Bauhinia, that it would be difficult to point out any group of plants in which there is more to instruct and delight the student.

The Pea tribe is so vast that the last enumeration of the species by Professor De Candolle, occupies between four and five hundred closely printed octavo pages. It will, therefore, be impossible for me to do more than give you a sketch of the general character by which this extensive assemblage is held together. *It consists of plants bearing pods*, formed upon the same plan as that of the Pea, and called Legumes ; this is the great essential character, and the only one which is universal. It is, therefore, necessary to teach you, in the first instance, how you are to know a Legume with certainty. Imagine to yourself a carpel which grows long and flat, and usually contains several seeds, and which, when ripe, separates into two valves or halves ; recollect, also, that the seeds all grow to one angle only of the inside of the carpel ;

in a word, study a Pea-pod, and you will know what a legume is. You must not expect, however, that it will always be exactly like a Pea-pod ; on the contrary, it is longer or shorter, larger or smaller, harder, thinner, or differently coloured, contains more or fewer seeds ; or, in short, may vary in many ways : but it will always be formed upon the same plan. This is what you are to take as the character which holds together all the subdivisions of the Pea tribe.

The most striking feature in these plants, next to the legume, is the singular arrangement of the petals, which gives to a very large proportion of the whole natural order the name of *Papilionaceous*, or Butterfly-flowered. By this title, we distinguish the first division of the Pea tribe ; as an example of which a common Pea flower would answer the purpose. I however, send you a sprig of the *narrow-leaved Restharrow* (Ononis angustifolia, Plate VIII. 2.). It has leaves the veins of which are at first sight ribbed rather than netted; you will, however, find, that the netted structure is what they really possess ; at their base is a pair of stipules as in the Roses and their allies. The calyx is formed of five sepals, that unite in a short tube *(figs. 2. & 3. a.)*. The corolla consists of five petals, one of which is larger, and stands at the back of all the others, wrapping them up before the flower expands *(figs. 2. & 3. b.)*; this is the *standard*, or vexillum. In front of the standard are two smaller petals *(figs. 2. & 3. c.)*, which are placed nearly parallel with each other, converging a little at the point ; they are the *wings*, or alæ, and

are carefully folded over a boat-shaped curved part of
the corolla, which is placed in front of all the rest;
this part, called the *keel*, or carina (*figs.* 2. & 3. *d.*),
is formed of two petals, which are slightly united at
their lower edge, as you may discern by pulling the
keel away from the calyx, when you will see their
two stalks (*fig.* 4.); the corolla is, therefore, formed
of the same number of parts as the calyx, but so
masqued that you would not have at first suspected
such a thing. This is what is called a butterfly-
shaped flower; some poetical Botanists having fancied a resemblance between the expanded flower and
a butterfly at rest.

Let us miss the stamens for the present, and pass
on to the ovary (*fig.* 6.), which is a tapering green
hairy part, gradually narrowing into a style which
ends in a minute stigma. Its legume is a short flat
body (*fig.* 7.), to which the withered style sticks.
When ripe it splits into two halves, to each of which
a seed or two (*fig.* 8.) is attached.

Papilionaceous flowers may be themselves separated
into those which have their *stamens united*, and those
which have their *stamens separate*. To the first belongs
the Restharrow; which has nine of the stamens joined
together about half-way (*fig.* 5.), and a tenth a little
separated from the others. It is here, also, that are
found nearly all those species of the Pea tribe with
which you are likely to be acquainted. *Peas, Beans,
Vetches, Clover, Trefoil,* and *Lucerne,* are known to
every body; all these you can easily procure for examination. *Laburnum,* too, with its branches of

golden flowers, *Furze* (Ulex europæus), and *Broom* (Spartium scoparium), which are almost too beautiful to be inhabitants of northern countries, *French Honeysuckles* (Hedysarum coronarium), with their crimson clusters ; and the singular *Bladder Senna* (Colutea arborescens), the pods of which explode with a loud report when smartly pressed between the fingers, are species of Papilionaceous plants of frequent occur-rence : all which it would be well for you to study, and compare with the characters you will find in systematic works. *Indigo* (Indigofera tinctoria), which has proved so valuable as a dye, you will not meet with easily in this country ; but you may procure the *Liquorice* plant (Glycyrrhiza glabra) the roots of which are so exceedingly sweet.

The second group of Papilionaceous flowers, with the stamens separate, comprehends the chief part of the showy New Holland shrubs, called *Pultenæa*, *Gompholobium*, *Daviesia*, &c., but not a single European species, nor any thing worth pointing out for its utility.

The next division of the Pea tribe consists of the *Cassias* and their allies, which are remarkable for *not* having Papilionaceous flowers ;—in place of which they have their petals spreading equally round the pistil, as in other plants ; their stamens are also usually spreading and separate. The irregularity of growth which causes the Papilionaceous appearance in the last division, also exists amongst these plants, so that you will generally find them with some of the petals or the stamens larger than the remainder.

Few of them are ever seen in this country, but in foreign climes they are exceedingly abundant. *Cassias* themselves, some of which yield the well-known medicine called *Senna*, are common in all parts of the tropics; the *Logwood*, the *Tamarind*, the *Barbadoes flower-fence* (Poinciana), the brilliancy of whose orange-coloured flowers is too intense to be steadily looked upon, the fragrant *Asoca Tree* of India (Jonesia), which Botanists have consecrated as a floral monument to one of the most learned of Oriental scholars, and the *Judas Tree* (Cercis siliquastrum), which makes all Turkey put on a violet robe, in its flowering season, belong to genera of the Cassia division. To these may be added, as other remarkable plants, the *horrid Acacias* (Gleditschias), whose trunks are covered with stiff branching spines, and which are so very remarkable in cold countries for the airy Mimosa-like appearance of their foliage; the *Carob* Tree, or *Algaroba* (Ceratonia Siliqua), the sweet pods of which are used for food in Spain, and whose seeds are supposed to have been the original *carat* weight of the goldsmiths, the *Tonga-bean* plant (Dipterix odorata), with the perfume of whose seeds you are doubtless acquainted; and, finally, Bauhinias, those large climbers which hang among the trees of tropical forests, like enormous cables, twisting round trunks and branches till they utterly destroy them.

The third division of the Pea tribe is that of *Mimosas*. Figure to yourself a plant with the sepals and petals of the Cassia group, only so small as

scarcely to be visible; the flowers growing in compact clusters; and the stamens not only very numerous, but so long and slender and delicate as to
resemble silken threads tipped with very little anthers.   If you can imagine such a structure you will
have a sufficiently correct idea of the Mimosa division, the beauty of which, on account of the multitude
of flowers they bear, and the gay colours in which
they are invested, is of so peculiar an appearance,
that one of them, a nearly hardy tree (Acacia Juli-
brissin), is actually called by the Persians, in whose
country it grows, the *Gul ebruschim*, or *Rose of Silk.*
Here belong the curious *Sensitive plants* (Mimosa),
whose many parted leaves shrink from the touch of
the very wind that blows upon them, which close up
and appear to go to sleep at night, and which seem
as if struck with sudden death if they receive any
rude shock; of these plants the balls of flowers are
peach-coloured.   To the genus *Acacia*, of which
great numbers are found in New Holland, where
they are called *Wattle Trees*, also belong the spiny
gum trees of Arabia and Senegal; the greater part of
the species have yellow flowers, and many of them
broad dilated leaf-stalks, in room of the many-parted
leaves which they bear when young.

The differences of these three divisions of the Pea
tribe may be expressed thus : —

Flowers papilionaceous . . *Papilionaceous Plants.*

Flowers not papilionaceous { Stamens few—*Cassias.*
{ Stamens numerous—*Mimosas.*

With this I take my leave of you for the present, promising that my next letter shall be shorter, if it is not more interesting.

EXPLANATION OF PLATE VIII.

I. THE ROSE TRIBE.—1. A leaf, a few flowers, and fruit of a *Strawberry Plant; a* stipules.—2. The calyx and pistil cut through to shew the origin of the stamens ; *a* sepals, *b—c* tube of the calyx.— 3. A stamen seen in front.—4. The same seen from the back.—5. A carpel ; *a* the ovary, *b* the style, *c* the stigma.—6. A fruit cut through perpendicularly to shew the fleshy receptacle, and the grains, *a*, sticking to it. Compare this with fig. 2.—7. A ripe grain.—8. The same cut through to shew the seed.—9. A seed extracted from the grain.—10. An embryo, with the radicle at the upper end.

II. THE PEA TRIBE.—1. A piece of the *Narrow-leaved Restharrow* (Ononis angustifolia).—2. A flower seen from the side ; *a* sepals, *b* standard, *c* wings, *d* keel.—3. The same flower seen in front ; the letters refer to the same parts.—4. A keel apart, shewing the two stalks of the petals which form it.—5. Stamens.—6. A pistil.—7. A ripe legume with the calyx adhering to it.—8. A seed ; *a* the cord by which it was attached to the receptacle.—9. The same cut open, shewing the position of the embryo, of which one cotyledon only is visible.

K

LETTER IX.

(Plate IX.)

If you cast your eye back over the various tribes we have passed in review, you will remark that they all agree in one circumstance, however different they may be otherwise. They all have both calyx and corolla : that is, two distinct rows of parts on the outside of the stamens ; and the petals are never joined together : on this account they are called *Polypetalous*, a name which is made by many Botanists to designate the large portion of the vegetable kingdom to which they belong. By and bye you will learn that another portion of considerable extent is called *Monopetalous*, because the petals are united by their edges into one tube or body ; and I have now to explain that a good many natural orders, which have either a calyx without any petals, or no calyx at all, receive the name of *Apetalous*. They are often also called imperfect or *incomplete*, with reference to their want of petals ; it is to some of these that I wish next to direct your study.

The first with which we shall begin, is what is called the *Protea tribe*, a group of exotic plants very much cultivated in green-houses for the sake of their beautiful or singular foliage, and the great masses of

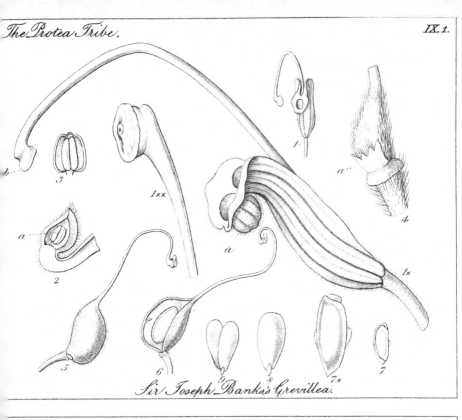

Sir Joseph Banks's Grevillea.

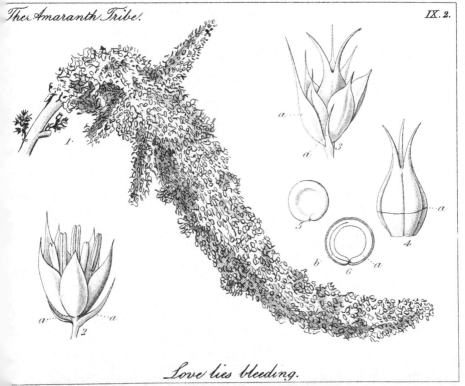

Love lies bleeding.

minute flowers which are borne by some of them ; but
totally unknown in a wild state in Europe. It would
be easy to name some kind with which you might
make a personal acquaintance, by inquiring of your
gardener after *Hakeas,* or *Persoonias,* or *Grevilleas ;*
but you will probably prefer that I should in this instance send you a copy of a drawing by Mr. Ferdinand Bauer of one of the handsomest of the kind
found by him in New Holland, and named *Sir Joseph
Banks's Grevillea* (Grevillea Banksii).

The leaves of this plant, like those of the whole
tribe, are exceedingly dry and hard ; they are
divided into many narrow lobes, but this is not by
any means universally the case, on the contrary they
are frequently perfectly simple and undivided.   The
calyx (Plate IX. 1.) is a long narrow tube, slit on one
side *(fig.* 1*.), and turned down at the point so as
to give the border a very oblique and bagged appearance ; by disturbing the bag with the point of
a pin it will divide into four concave lobes, each of
which *(fig.* 2.) allows an anther to nestle within its
cavity. The pistil consists of a long hard style, rather
abruptly bent above the middle, terminated by a
thickened one-sided stigma *(fig.* 1**. & 1*. *b.*), and
arising from a hairy one-celled ovary having a jagged
scale at its base *(fig.* 4. *a.*).   This scale is one of
the things which used to be called *nectary,* under the
idea that it was formed for the purpose of secreting
honey or nectar ; but that term is now abandoned.
The style is so long that you would wonder how
it could ever have been confined within the calyx ;

and the stigma is so far off the anthers that you will
find it yet more difficult to imagine how the pollen is
to touch it; both these are arranged in a very simple
way.   Before the flower opens, the stigma, as the
style lengthens, is pressed against the point of the
calyx; but here the sepals adhere so firmly that
they will not open ; the consequence of which is that
as the style goes on lengthening it gradually takes a
bend upwards, and pressing forcibly against the
upper side of the calyx, splits it open, by separating
the two sepals upon whose line of union it is forced.
The pressure of the stigma upon the point of the
calyx causes the latter to be moulded into a sort of
socket, in which the anthers actually lie ; so that as
soon as the stigma begins to be loosened, by the
growth of the style after the latter has slit the calyx,
the pollen is gently taken out of the anthers by the
cup of the stigma ; which, when it finally separates
altogether and rises up, carries the pollen away along
with it.

In time the calyx falls off, and the ovary grows
into a hard dry fruit (*figs.* 5. 6.), which opens like
a legume, and exposes to view a couple of seeds.

Other Proteaceous plants are formed upon a similar
plan : their calyx is often separated into four distinct
sepals, and then no socket is formed to hold back the
stigma ; or there are other variations of minor importance, but in the absence of petals, in the origin
of stamens from the face of the sepals, and in the
peculiar fruit, they all agree.

What causes the most striking difference in their

appearance, is the flowers of some growing singly among the leaves, and of others being collected into compact heads. Those genera which have the latter structure, are the handsomest and most usually cultivated ; the Proteas which are found at the Cape of Good Hope, in dry, barren, stony, exposed situations, are most noble looking objects, in consequence of their beautiful feathery flowers being half hidden by large red, or white, or black-edged bracts of the purest colours. Banksias and Dryandras are chiefly valued for their handsome leaves ; some of the latter are so fringed with long hairs as to resemble the plumes of birds. They are applied to scarcely any useful purpose ; but appear to be perfectly harmless : their seeds are sweet, and are eaten sometimes as nuts, both in Africa and South America ; one of them, called *Witteboom* (Protea argentea), is the common fire-wood at the Cape of Good Hope.

But let us leave these showy and useless strangers for a tribe that is known to every one who has a garden. *Love-lies-bleeding* (Amaranthus caudatus), *Prince's Feathers* (Amaranthus), *Globe Amaranths* (Gomphrena globosa), *Tricolors* (Amaranthus tricolor), and *Cockscombs* (Celosia coccinea), have been cultivated as long as gardens have been prized ; they form, along with some others of a similar structure, what is called the *Amaranth tribe*. This natural order, like the last, has no corolla. Its calyx consists of five crimson sepals (*figs.* 2. & 3.), of so dry a texture, that you would say they were really dead ; these are surrounded by a number of bracts, of the same colour and texture as themselves. It is owing to the dryness and thinness, and usually gay colours of these parts, that Cockscombs and the like owe their glossiness and beauty, and also the property they possess of remaining for months without fading. The remainder of their organs are constructed upon the simplest plan. A few anthers, usually five (*fig.* 2.), and an ovary with two or three styles (*fig.* 4.), having but one cell and one ovule, complete the apparatus by which such a plant is increased. When the fruit is ripe, the shell of the ovary becomes very thin, and bursts in the middle by a horizontal opening (*fig.* 4. *a.*); the seed (*fig.* 5.) is a little flat body, with a slender embryo (*fig.* 6.), coiled round some mealy albumen.

It is difficult to mention an order much more simply constructed than this, and yet how perfectly are all the parts adapted to the end for which they are created. Even a provision for a beautiful appearance is not neglected, for in order to compensate for their smallness, we find the flowers developed in large masses, and aided by multitudes of shining bracts which contribute very essentially to their fine appearance.

With an assurance that these plants are all as harmless as they are beautiful, I take my leave of you till another day.

EXPLANATION OF PLATE IX.

I. The Protea Tribe.—1. A flower of *Sir Joseph Banks's Grevillea*, seen in front, of the natural size.—1*. The same magnified and viewed from the sdie ; *a* the socket of the calyx, *b* the stigma.—1**. The upper end of the style and the stigma viewed in half profile.—2. The upper end of a sepal, with the anther *a*, nestling in it.—3. An anther.—4. An ovary, with the scale, *a*, at its base.—5. A ripe fruit, natural size.—6. The same burst open.—7. The seeds taken out, with a moveable partition that separates them just brought into view. —7*. The same magnified.—8. An embryo.—9. The same with the cotyledons divided a little. (All these are after a figure by Mr. Ferdinand Bauer.)

II. The Amaranth Tribe.—1. A bit of the inflorescence of *Love-lies-bleeding*, of the natural size.—2. A calyx containing stamens : *a* bracts.—3. A calyx containing a pistil; *a* bracts.—4. A ripe fruit ; *a* the horizontal line where it opens.—5. A seed.—6. The same cut perpendicularly ; *a* the radicle, and *b* the cotyledons of the embryo.

LETTER X.

(Plate X.)

So accustomed are people to identify gay colours with the corolla of a flower, that it is always difficult to make them believe the ovary and red striped part of the Marvel of Peru, to be really a calyx. Such, however, is undoubtedly the fact.

The *Marvel of Peru* (Mirabilis Jalapa) is the representative of a *tribe* named after it, belonging, like the subjects of my last letter, to the Apetalous division of Dicotyledonous plants. It has fleshy perennial roots; jointed stems, which perish at the first attack of frost; and broad opposite leaves, with netted veins. Its flowers appear in compact clusters; and are each surrounded at the base by a green involucre, divided into five segments, so as to resemble a calyx, for which it would be certainly mistaken, if it did not sometimes bear two flowers within the same involucre; a kind of structure that never is found in a true calyx.

Each flower consists, firstly, of a funnel-shaped calyx (Plate X. 1. *fig.* I.), divided at the end into five orange and red plaited lobes, and contracted at

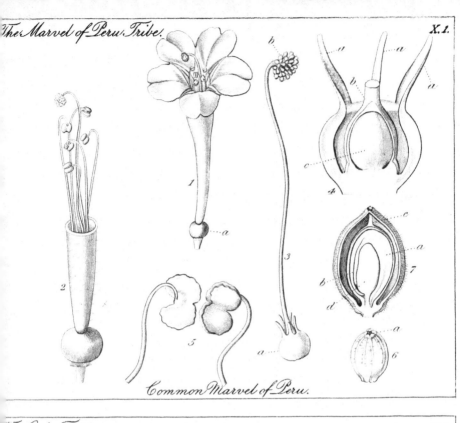

X. 1.

Common Marvel of Peru.

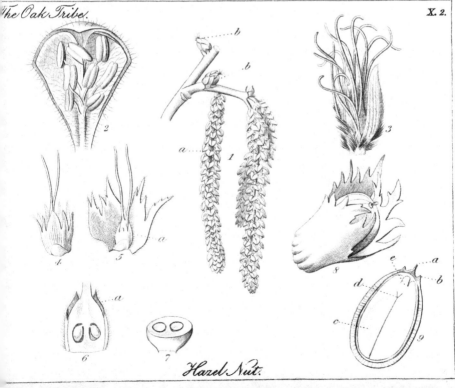

X. 2.

Hazel Nut.

the base (*fig.* 1. *a.*) into a roundish fleshy ball;
secondly, of five stamens of unequal lengths, arising
from below the ovary (*fig.* 3. *a.*) round which they
form a fleshy cup, and then adhering to the sides of
the calyx; so that they are actually perigynous and
hypogynous at the same time.

The ovary (*fig.* 4. *c.*) is a superior body, containing
a single ovule, which grows from the bottom of the
cavity; it is terminated by a long thread-shaped
style, which ends in a cluster of little round warts
(*fig.* 3. *b.*), forming a stigma. Thus far the structure
is as simple as that of the Amaranth and Protea
tribes; nor will it be found more complex in the
fruit.

As soon as the flower begins to fade, the roundish
fleshy ball at the bottom of the calyx, swells and
grows harder, contracting at the top, and in time
throwing off the thin and coloured part. At last it
acquires a woody texture, shrivels round the veins,
and becomes an oblong brown nut (*fig.* 6.), with a
little hole at its point (*a.*), where the upper coloured
part of the calyx fell off. Upon opening it, you will
find the fruit, with a very thin shell, and the remains
of the style at its top (*fig.* 7. *c.*). Within it lies a
single seed having an embryo (*fig.* 7. *a. b.*), rolled
round a quantity of mealy substance, which is the
albumen (*d.*).

Such is the character of the natural order that
contains the Marvel of Peru, which is by far the most
handsome genus it comprehends. Generally, the
order consists of obscure weeds, which are rarely

seen in gardens, although they are common enough
in tropical countries. It differs obviously from the
Protea tribe in the stamens being hypogynous; from
the Amaranth tribe in the calyx being all in one
piece; and from both in the singular circumstance of
the lower part of the calyx becoming hardened and
forming a sort of spurious shell to the fruit. This
last is the essential character of the Marvel of Peru
tribe, or Nyctagineæ; which I only introduce to you as
a striking instance, firstly, of the highly coloured
condition often assumed by the calyx, and secondly, of
the singular manner in which one part is occasionally
employed by nature to perform the part of another.

Very different from these, although also belonging
to the Apetalous division of Dicotyledonous plants, is
that most interesting natural order, which includes
the *Oak*, and the *Sweet Chesnut*, the *Beech*, the
Hornbeam, and the *Hazel;* in short, the larger part
of our common European trees. In consequence of
its containing the Oak it bears the name of the *Oak
tribe*. Until I shall have explained to you the real
origin of all the parts you find in these plants, and
the singular manner in which they change between
the infancy of their flowers, and their old age, you
will have had but a feeble idea of the wonderful
power the parts of plants possess of assuming unusual forms after they have been once developed. If
it be true that flowers are generally seen in a masquerade dress, as some Botanists poetically assert, it
certainly is here that their disguise is the most impenetrable.

The *Hazel* is one of the most accessible to you when it is young, and a good illustration of the structure of the others. At the earliest period of the spring you must have remarked the branches of the Hazel loaded with little yellow tails, which swing about as the wind disturbs them, and fill the air with a fine and buoyant powder, the particles of which may be seen glittering in the sunbeams like motes of gold. These tails are called CATKINS (Plate X. 2. *fig*. 1. *a*.), and are composed of a great number of little scales, which are arranged one behind the other with the utmost regularity, as you may easily discover by inspecting them, before they separate. Each scale has on its inner face about eight anthers, that seem to arise out of a two-lobed flat body, which adheres to the scale (*fig*. 2.); no other structure is to be found; apparently neither calyx, nor corolla, nor pistil; nothing but the two-lobed body sticking to the scale and bearing the stamens. Botanists consider the scales bracts, and the two-lobed body a calyx in an imperfect state.

This then is an instance of a simpler kind of organization, than any you have before met with in a flower. It is, however, not quite characteristic of the Oak tribe, for the Hornbeam has no calyx whatever, while the Oak, and the Beech, and the Sweet Chesnut have a much more perfect one than the Hazel.

If the Hazel had none but stamen-bearing flowers, you would never have any nuts in the autumn; for there is nothing in those flowers which could by any possibility change into a nut. In this plant not only are the stamens and pistils in different flowers, but in different parts of the plant, and organized upon quite a different plan. If you observe attentively those buds of the hazel which grow near the catkins (*fig.* 1. *b. b.*), about the time when the stamens are shedding their pollen, you will perceive some little red threads protruding beyond the points of the buds, and spreading away from the centre ; those are the stigmas, and the pistils are enclosed within their scales, where they are safely protected from accident and cold. At the earliest moment when the stigmas can be discovered, let the scales be removed (*fig.* 3.), and you will find the flowers clustered together among a quantity of soft hair, which seems provided as an additional means of shielding them from the weather, and to serve the same purpose as the warm lining of down, which the birds provide for their young when they first break the shell and before they are fledged. Each of these flowers is surrounded by a jagged sort of cup (*fig.* 4 & 5.), which is originally much shorter than they are, but which in time grows considerably longer ; that cup is the involucre. The flower itself consists of a jagged superior calyx (*fig.* 5. *a.*) ; an ovary with two cells and two seeds (*fig.* 6.), and two long thread-shaped crimson stigmas. Thus you see the calyx of the pistil-bearing flower is much more perfect than that of the other kind of flower ; but it is still very imperfect.

The pistils and the stamens being thus separated, there would be no chance of the pollen of the one falling on the stigma of the other, and fertilizing it, unless an unusual quantity of stamens was provided; hence it is that in a fine day in spring the whole air is, as I have just said, so impregnated with particles of pollen, that they cover every thing as with a fine dust.

By degrees, as warm weather advances, the protection of the scales of the bud is no longer necessary to the young flowers, which swell and burst through them ; the involucre daily grows larger ; the stigmas having fulfilled their destiny, shrivel up ; the ovary enlarges ; one of its ovules grows much faster than the other, and gradually presses upon it till it smothers it ; the shell hardens, an embryo makes its appearance, and by degrees fills up the cavity ; and at last you have a perfect nut, with its husk (*fig.* 8.), or involucre. At the point of the nut is to be seen the remains of the calyx (*fig.* 9. *b.*) ; but no trace can be found of the cell and ovule which were smothered ; so that a one-celled fruit is produced by a two-celled ovary. You will now know why nuts sometimes grow in clusters, and sometimes singly ; if cold or accident should destroy any part of the cluster of pistils in the bud, but a very few nuts, perhaps only one, will grow and ripen ; but if they are mostly saved, you will then have the large clusters which are so common in seasons which have been preceded by mild springs. The nut itself affords an excellent illustration of the structure of a dicotyledonous embryo ; the two great fleshy lobes into which the nut separates when freed from its skin (*fig.* 9. *c.*),

are the cotyledons ; the little conical part at one end
(*e.*) is the radicle, and the small scale-like body which
lies between them in the inside (*d.*) is the plumule,
or young stem.

Still more curious than those of the Hazel are the
changes that occur during the growth of the fruit of
other genera of the Oak tribe.   In the *Oak* itself the
involucre is formed of a great many rows of scales,
which gradually grow larger and harder, and more
numerous, and at last become what you call the cup
of the acorn ; a part you would never have guessed
could have been made out of a number of little leaves,
if you had not watched their successive changes.
The ovary at first contains three cells, and each cell
two young seeds ; but in obedience to the constant
command of nature, one of the seeds grows faster
than the rest, presses upon the other cells and seeds,
gradually crushes them, till at last, when the acorn
is ripe, all trace of them has disappeared.

In the *Beech*, the involucre originally consists of a
vast quantity of little thread-like leaves, which en-
close a couple of pistils. These leaves gradually grow
together, and over the pistils, so as to form a prickly
hollow case, which completely encloses the nuts ; at
last, the case rends open spontaneously into three or
four woody pieces, and makes room for the nuts, or
mast, to fall out.   As in the Oak, one of the ovules
destroys all the others, so that out of six young
seeds, but one is found in the ripe nut ; here, however, you may generally find the five that have pe-

rished remaining like little brown specks, sticking to the top of the cell of the nut.

In the *Sweet Chesnut*, alterations in character, and the destruction of one thing by another are carried still further. In that plant, the involucre, which, when full-grown, is a hollow case, covered over with rigid spines, was in the beginning a number of little leaves, which gradually grew together as in the Beech ; they kept acquiring with their age a greater degree of rigidity ; their veins separated, and formed clusters of spines, till at last the whole surface of the husk was covered with little spiny stars ; each star was in the beginning a leaf, and its rays the veins of the leaf. The pistils each contained six or seven cells, with a couple of ovules in each ; yet the ripe nut has only one seed : so that in the course of the growth of a chesnut, no fewer than six cells, and thirteen ovules are destroyed by the seed which actually grows.

I feel sure you will now agree with me, that if any plants can be said to exhibit themselves in a masquerade dress, these are they. For without this explanation, who could have supposed that the husk of the filbert, of the beech, and of the chesnut, were all of the same nature, and constructed upon the same plan as the cup of the acorn ; and especially, who could have supposed that the chesnut, with its single seed, could ever have originated from an ovary of seven cells and fourteen ovules.

It is among these trees that you will find the best specimens of the Exogenous structure of the wood of

Dicotyledonous plants, a subject upon which I as yet have said almost nothing. Get a branch of Hazel, Beech, or Oak, and divide it horizontally, so as to have a view of the whole of the inside from one side to the other, the section forming a circle which is bounded by the bark. Let the section be rendered smooth by a sharp knife or a plane, so that all the parts may be distinctly seen; you will then remark in the centre a pale roundish spot, which a glass will shew to be formed of a soft spongy substance: it is the PITH, a cellular provision of nature for the support of the young buds, when they are too weak to obtain food from more distant sources. Next the pith follow several rings of WOOD, each of which is composed of an infinite number of tubes, and was the produce of one year's growth, so that the number of the rings tells you the number of years the branch has been in acquiring its present size; the most external of the rings is the youngest, and also the palest, while the most internal or the oldest is of a deep brown colour; the pale is called the SAPWOOD, the brown the HEART-WOOD; the latter, which is much more durable than the former, is filled with a substance of a hardening nature, which was originally formed in the leaves, and which is stored up by Providence in the centre of the stem, where it lies beyond the reach of accident or injury, until old age comes and produces decay; originally the heartwood was sapwood, and that which is sapwood now, would have become heart-wood in a few years, if you had not cut off the branch. It is in the sapwood chiefly that the vital

energy of the stem resides, and it is through it that
the sap rises in the spring for the supply of the buds
and leaves ; an Exogenous tree can therefore lose the
whole of its inside without suffering much diminution
of growth, so long as the sapwood remains uninjured,
and this is the reason why trees that are hollow go on
growing century after century, just as if their inside
were still sound.   On the outside of the wood is the
bark, which binds up and protects all the other parts,
and down which the returning current of sap descends
towards the roots.

Next carry your eye attentively over the section,
from the bark to the pith, and you will remark that a
number of fine pale lines are drawn as it were from
one to the other, forming delicate but broken rays ;
these lines, which are composed of flattened cells,
and named MEDULLARY RAYS, are in reality the ends
of extremely thin plates which connect the pith and
the bark together ; they perform an important part
in the system of vegetation, for it is they which convey the descending sap from the bark to the centre
of the stem ; it therefore is they which are the cause
of the production of heartwood, and all trees without
them must be destitute of it; as is the case with
monocotyledonous plants, which are always softest
in the centre.

Thus you see the sap, which rises from the roots is
carried upwards in the sapwood, down again in the
bark, and laterally into the hidden recesses of the
trunk, by the medullary rays, all three currents

L

moving at the same instant, but without the slightest
interference with each other.

Can any thing be more admirably adjusted that all
this? By means of a system of tubes and cells va-
riously arranged, the whole of the important business
of the conveyance of food to the leaves, and of the
peculiar properties formed there from the leaves to
the centre of the stem, and down to the extremest
roots, is carried on in the most certain and effectual
manner, even in the loftiest and most gigantic forest
trees. Only conceive what a wonderful combination
of powers must be provided to enable a tiny leaf, not
perhaps half an inch long, on the highest branch of
a tree, to procure its food from roots sometimes 250
feet off, or at a distance equal to six thousand times
its own length.

But I must not dwell further upon this subject, interesting as it is ; to works on Vegetable Physiology
you must refer for a full elucidation of all such
matters.

Bearing in mind, that the Oak tribe is constantly
known by its imperfect apetalous flowers, and its
singular involucre, you are in no danger of forgetting
it. Formerly, the *Birch* and *Alder*, the *Poplar* and
Willow, and the *Plane*, were considered as also belonging to it ; and the whole were called *Amentaceous*,
because of their stamen-bearing flowers being constantly arranged in catkins, or *amenta*, as they are
technically designated. But these trees are so extremely dissimilar, in other respects, that they now
form several independent tribes.

Of the distinctions of these I shall by and bye give you some account; for the present, I must leave you till you have examined for yourself, as you easily may, the highly singular phenomena I have explained to you.

EXPLANATION OF PLATE X.

I. THE MARVEL OF PERU TRIBE.—1. A flower of the *Common Marvel of Peru* (Mirabilis Jalapa); *a* the thickened base of the calyx. —2. The same part magnified, with the upper part of the calyx cut away.—3. The fleshy base of the stamens *a*, from within which rises the thread-shaped style, terminated by the stigma *b*.—4. A perpendicular section of the fleshy base of the stamens and pistil; *a* the base of the separate parts of the filaments; *b* the base of the style; *c* the ovule seen in consequence of a part of the shell of the ovary being cut away.—5. Anthers.—6. A ripe nut; *a* the closed up orifice of the calyx.—7. A perpendicular section of the nut, shewing the fruit standing erect in the inside of the hardened base of the calyx; *c* the base of the style; *a* the radicle, and *b* the cotyledons of the embryo rolled round the mealy albumen *d*.

II. THE OAK TRIBE.—1. A twig of *Hazel* (Corylus Avellana); *a* the stamen-bearing catkins; *b* the buds containing the pistils.—2. A scale of the catkin, shewing the two-lobed body, and the stamens.—3. A cluster of pistils bearing flowers, in a very young state, with only one of the scales by which they are protected remaining.—4. A pistil-bearing flower, inclosed in its involucre.—5. The same cut open; *a* the calyx.—6. An ovary divided perpendicularly; *a* the calyx.—7. The same divided horizontally —8. A ripe nut in its husk, or involucre.— 9. A nut cut through perpendicularly; *a* the remains of style; *b* remains of calyx; *c* cotyledons; *d* plumula : *e* radicle.

LETTER XI.

THE NETTLE TRIBE—WOODY FIBRE—THE BREADFRUIT
TRIBE—THE WILLOW TRIBE.

(Plate XI.)

THE tribes of Dicotyledonous Plants, with only a
calyx to their flower, are far from being exhausted
with those we have already seen. On the contrary,
there is a great many races, the structure of which is
extremely curious. I must, however, be content
with selecting three of the commonest for further
illustration, and with referring you for a knowledge
of the remainder to the systematic works of Botanists.

Let one of these be the *Nettle tribe ;* we will not,
however, select the Nettle itself, because of its stings,
but begin with a harmless plant called *Common Pellitory* (Parietaria officinalis), which you may find
every where on old walls, or in dry waste places.
This species grows either prostrate, or erect, is very
much branched, has reddish stems and leaves, and
clusters of minute reddish-green flowers in the bosom
of the leaves (Plate XI. *fig.* 1.). Over all its surface
are scattered stiffish hairs, which do not sting, but

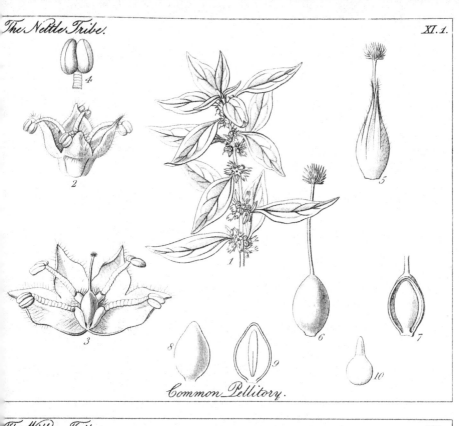

Common Pellitory.

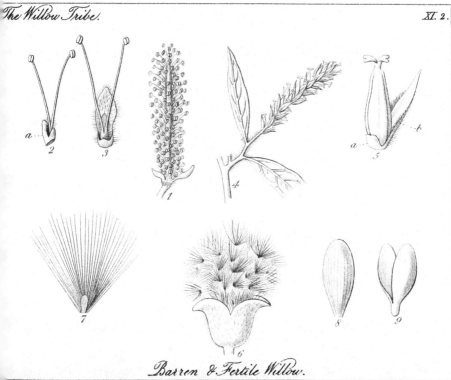

Barren & Fertile Willow.

which would if they were stiff enough. At the base
of each leaf is a pair of shrivelled brownish stipules.

The flowers of this plant are of three sorts, those
which have stamens only, those which have stamens
and a pistil, and those which have a pistil only. As
the first and second are constructed alike, let us
consider them as essentially the same ; especially as
their pistil is seldom perfect. To each of these flowers
there is a calyx with four hairy divisions (*fig.* 2 & 3.);
opposite to each division is a stamen; and in the centre
is a pistil more or less imperfect. The filaments are
worthy of examination ; their lower end is firm,
smooth and fleshy; but it abruptly alters to a
withered shrivelled part, so dissimilar in aspect that
you would think it must be a distinct organ. Before
the flower opens the shrivelled part is pressed down
by the segments of the calyx, which are finally forced
asunder by the filaments with force, so that the
flower opens with some degree of elasticity ; a provision, in all probability, to secure the scattering
of the pollen, by which the distant pistils may be
reached.

The flowers that contain the pistils are mixed
among the others, and like them consist of a calyx
with four divisions ; but as this contains no stamens,
its figure is not roundish, but sharp-pointed, like
that of the ovary, to the surface of which it is closely
applied. The pistil consists of one ovary containing
a single seed (*fig.* 7.), of a thread-like style, and of
a pin-headed stigma, the little fringes of which

spread in all directions, and are admirably contrived
to catch the grains of pollen floating in the air.

The fruit is an oval shining blackish lenticular grain
(*fig.* 8.), which contains a single seed, with an inverted embryo lying in the midst of fleshy albumen
(*fig.* 9.).

Such is essentially the manner in which all the
remainder of the Nettle tribe is organized. *Pistils
and stamens in different flowers, leaves covered with
rough or stinging hairs, elastic stamens, and lenticular
grains,* are common to them all.

Nettles, which are so remarkable for the intolerable
pain, and even the sometimes highly dangerous effects caused by their stinging hairs, differ from Pellitory chiefly in their pistil-bearing flowers having a
calyx of two sepals.

Hops have not only a twining stem, and their
pistil-bearing flowers collected in leafy heads, but
are also known by having five stamens in each sterile
flower, and the pistils and stamens on different plants.

Finally, *Hemp,* which also belongs to the Nettle
tribe, has the calyx of the pistil-bearing flowers slit
on one side, two unequal styles, and five stamens in
the sterile flowers. The peculiar tenacity of the stems
of Hemp is not uncommon in other plants of the same
natural order, and may even be considered characteristic of it.

Considering the many important purposes to
which the Hemp is applied, and the prodigious
strength that it possesses when twisted into ropes,
you will probably be curious to know something of

the exact nature of the part which is capable of being converted to these great purposes. In that case you must have recourse to your microscope; beneath which you may place a little tow, the threads of which are separated, and float in water. At first sight, with a weak magnifying-glass, you will discern no distinct organization in these threads; they will look like dark lines of about the thickness of a fine human hair; but if you bruise them and tear them about in the water with the point of a couple of needles, you will in time succeed in separating each of the threads into a very considerable number of exceedingly fine parts, which you may discern, by increasing the magnifying power of your microscope, to be transparent tubes, composed of a tough membrane, and tapering to each extremity like bristles; these are glued together in bundles which constitute the finest threads that are visible to the naked eye. Their business is not simply to grow in the inside of a plant, in order that man may pull them out and apply them to his own purposes; they have a far higher and more important office to fulfil. It is they which give strength and toughness to every part, and which enable the stem and the leaf to wave about in the breeze or to bend before the storm without breaking; they are placed as a sort of sheath all round such tender parts as the spiral vessels, which are enabled within their safeguard to perform their delicate functions with certainty and security; and, finally, it is they which act as so many water-pipes to convey the fluids of plants with rapidity from one part to another ; up the wood, along the veins of the leaves to their extremest points, back again into the stem, and down the bark towards the roots. When plants have no woody fibre, they are universally so delicate and weak as to be unable either to raise themselves in the air, or to withstand any violence, as we see in mosses, lichens, mushrooms, and such plants.

You will probably be surprised when I tell you that the *Fig Tree* is so nearly the same in structure as the Nettle, that many Botanists consider it to belong to the same tribe ; and you may possibly be tempted to exclaim with some who have not considered the subject very attentively, " How absurd to place the Nettle and the Fig together in the same natural group !" *I* must admit that this does appear strange until the reason is pointed out ; and I trust *you* will admit that it is clearly right when the reason shall have been explained to you.

Let us then see what a Fig Tree is. It is an Exogenous plant, with leaves covered with very stiff short hairs, and with a pair of stipules at their base ; so is a Nettle. It has flowers with stamens and pistils separate ; so has a Nettle ; its flowers have no corolla, and the pistil is a little simple body, which changes, when ripe, to a very small flat grain ; all which is exactly what we find in the Nettle. In the essential parts of their structure the two plants then are alike. But where are the flowers of the Fig, you will inquire ; you can see nothing but a thick oval green body, which you know will turn to fruit, and which therefore ought to be the flower ; here, however, you must again be prepared to meet with natural wonders. The thick oval green body is a hollow box, or receptacle; within it in darkness and obscurity are reared the flowers, which, like the beggars' children in the caverns among the fortifications of Lille, are so deformed and pallid as hardly to be recognised. Cut a young fig open; the whole of its inside is bristling with sterile and fertile flowers, the former having five stamens, and the latter a jagged calyx, with a little white pistil sticking up in the midst of it. This pistil, when ripe, becomes a flat round brown grain, which is lost among the pulp of the fleshy and juicy receptacle, where you eat it, and call it a seed.

The difference then between the Nettle and the Fig Tree consists not in the structure of the stem, or of the leaves, or of the calyx, or stamens, or pistils, or fruit properly so called; but in the hollow fleshy receptacle within which the flowers are forced to pass through their different stages. This kind of difference is, however, of a very unimportant kind; and not greater than you find between the Strawberry and the Rose, about whose relation to each other every one is agreed.

For these reasons, both the Fig and the Nettle are by some considered to belong to the same natural order; there is, however, a difference that I have not mentioned, and which is important; the juice of the Nettle is watery, while that of the Fig is milky; on which account other Botanists consider the Fig to be the representative of a natural order, distinct from that of the Nettle, but in the closest affinity with it.

To this natural order belong the *Breadfruit tree* (after which it is called the *Breadfruit tribe*), the *Mulberry*, and many other exotic trees. For their milkiness they all are most remarkable; it is usually of a somewhat acrid nature, as you may find in the Fig itself; sometimes is highly poisonous as in the *Upas tree* of Java and some Indian species of Fig; or is again quite harmless and even nutritious in the *Cow tree* of South America, to the trunks of which, the Indians repair in the morning with their jugs and pails, just as the milkmaids of Europe to their cows. It is, however, probable that in this instance the milk is harmless only at a certain period of the year, before the venomous principle is formed; for a West Indian plant called Brosimum, the young shoots of which afford a wholesome food for cattle, is very nearly the same as the Cow Tree, and its old shoots are poisonous. The Fig itself would not be fit to eat if gathered green, because at that time the fruit abounds in milk; but when it is ripe all the milk has dispersed, and then only it becomes the wholesome fruit with which we are so well acquainted.

Let this be a lesson to you never to judge hastily of the affinities of plants, but to remember that it is structure alone, and not vague external resemblances or differences by which their relations are determined botanically.

The last of the tribes without corolla, which I shall be able to lay before you, is one that is constructed with still more simplicity than the last. They had at least a calyx, but this has neither calyx nor corolla,

nor any sort of covering to the stamens, beyond the
scale like bracts, out of the bosom of which the stamens or the pistils arise. These plants are *Poplars*
and *Willows*, which together form the *Willow tribe*
(Plate XI. 2.). Their flowers grow in catkins (*figs.*
1 & 4.)—those beautiful silky bodies, glittering as
it were with gold and silver, which are hailed by
northern nations as the earliest harbingers of spring,
and gathered for festivals under the name of *Palms*
near London, and of *gostlings* in some parts of England. The stamens are upon one plant, the
pistils upon another. The former are one, or two, or
three, or five, or more, to each bract (*figs.* 2. 3.); the
latter are seated singly within a bract, and consist of an
ovary, having one cell with many seeds, and a lobed
stigma (*fig.* 5.). The fruit consists of hollow cases,
which split into two valves (*fig.* 6.), and discharge a
multitude of small seeds, covered with fine hair or
wool (*fig.* 7.), like the seeds of the cotton plant. On
these downy pinions the seeds will fly to great distances and scatter themselves over the whole face of
the country. The Willow is absolutely without any
trace of calyx; the Poplar has a sort of membranous
cup, which may be considered the rudiment of one.

In taking leave of these imperfectly formed orders,
I would recommend you to reduce their characters to
an analytical form, in order to see their differences the
more distinctly. You may do this in many ways,
of which the following will serve as an example.

Two of them have the sepals combined into a tubular calyx, namely, the Protea and Marvel of Peru

tribes; of which the former has hard leaves and stamens placed opposite the sepals, and the latter soft leaves, with the bottom of the calyx forming a bony covering to the ripe fruit.

Two others have their male flowers arranged in catkins, namely the Oak and the Willow tribes, of which the former has closed fruit seated in an involucre or cup, and the latter opening fruit without any involucre.

Finally, the three that are remaining, namely, the Amaranth, Nettle, and Breadfruit tribes, are at once known by the first having smooth leaves without stipules, while the two last have rough or stinging leaves with stipules.

These peculiarities may be expressed in a tabular form thus :—

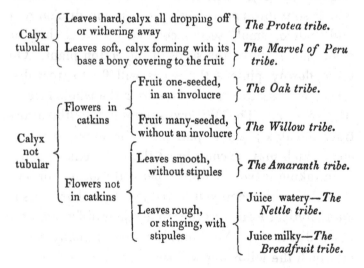

Our next visit will be to far more beautiful subjects.

EXPLANATION OF PLATE XI.

I. THE NETTLE TRIBE.—1. A piece of *Common Pellitory* (Parietaria officinalis).—2. The calyx of a sterile flower, with the stamens
projecting.—3. The same cut open.—4. An anther.—5. A fertile
flower.—6. A pistil.—7. An ovary cut through perpendicularly,
shewing the position of the ovule.—8. A ripe fruit.—9. The same cut
through perpendicularly, shewing the position of the embryo within
the albumen.—10. An embryo.

II. THE WILLOW TRIBE.—1. A catkin of sterile flowers of the
Monandrous Willow.—2. A single sterile flower of the *Yellow Osier*
(Salix vitellina), with the gland, *a*, at its base.—3. The same with the
bract that belongs to it.—4. A catkin of the fertile flowers of the *Yellow
Osier.*—5. The pistil with its gland, *a*, and its bract, *b.*—6. A seed-vessel
ejecting the seeds.—7. A seed with its down.—8. A seed without the
down.—9. An embryo with the cotyledons separated.

LETTER XII.

THE HEATH TRIBE—THE BINDWEED TRIBE.

(Plate XII.)

LET me now introduce you to an extremely beautiful set of plants, in which you will find nothing but the most charming colours, set off by so clear a complexion, and such perfect forms, that there is little comparable to them in the whole vegetable kingdom.

Very different from the tribes we have lately seen, so far are they from wanting a corolla that they possess that part in a most highly expanded state ; not, however, consisting of several distinct petals, as in all the natural orders we at first examined, but having the petals grown together into a cup or bell, or hollow body of some kind, and only separate at their upper ends. These corollas are technically named *monopetalous*, or one-petaled ; a very bad designation, because the corolla consists in reality of many petals in a united state ; but the word was invented when the real nature of such a corolla was unknown, and custom has established what error first promulgated. It is to the various tribes of *Monopetalous Dicotyledons* that we are now to direct our attention.

The *Heath tribe*, which is what I at first alluded to, stands pre-eminent among such plants for its love-

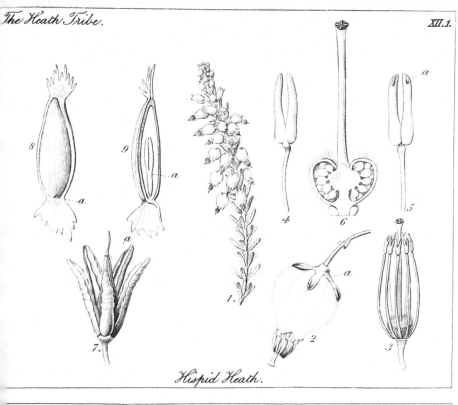

Hispid Heath.

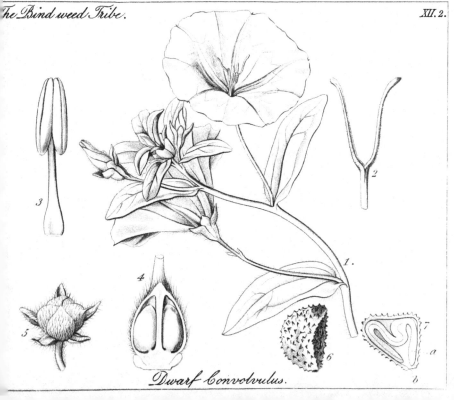

Dwarf Convolvulus.

liness; and it is very easily known. We will say nothing about its leaves, firstly, because they are variable in appearance; and, secondly, because it is now time we should leave off testing the Dicotyledonous character of a plant at every step; for you must by this time have begun to recognize, with certainty, the features of that primary class, without attention to its technical distinctions. It is in the flowers that the great peculiarities of the tribe are conspicuous. Let us take any Heath for an examination; that which I have at hand is the " Hispid" (Erica hispida), so called on account of the little stiffish hairs with which it is covered on every part. Like nearly the whole of the genus, it is a native of the Cape of Good Hope, where extensive tracts are covered with countless species of all manner of forms and colours. The Hispid Heath has a calyx of four sepals (Plate XII. 1. *fig.* 2. *a.*); and a corolla looking like a hollow globe, with four short teeth at one end; it is of the clearest pink; you may see its veins through the skin, so transparent is every part; and their arrangement will reveal to you the fact of this hollow globe being in reality composed of four petals, so completely united at their edges that nothing but their points is to be distinguished.

Arising from beneath the ovary, and perfectly separate from the corolla, are eight stamens (*fig.* 3.), each of which has a slender filament, and a singular purple anther, with two distinct lobes, shaped like the two prongs of a fork, and opening by a hole at their upper end. This character of holes in the end of the anther is one of those that are essential to the Heath tribe.

The ovary (*fig.* 6.) is a hairy body containing four cells and a great many ovules ; it is terminated by a style having a flat purple stigma, with four little projections on it, corresponding with the number of cells in the ovary. This in time changes to a dry fruit that bursts into valves, for the escape of a countless multitude of seeds as fine as grains of sand ; they are frequently terminated by delicate crests or wings of different figures (*figs.* 8. & 9.), and are beautiful microscopic objects.

Now in this description you are to consider the *hypogynous stamens, and the anthers with pores in them*, as the most essential characters of the natural order, and they will, in fact, enable you to distinguish it from all others. It is principally in the breadth of the leaves, in the size and form of the flowers, in the texture of the fruit, and in the number of divisions of the corolla and stamens that the genera vary : they all agree in those common characters.

For instance, the *Arbutus* is like a Heath, but it has broad leaves, ten stamens, and fleshy fruit, which gives rise to its common name of the *Strawberry tree*, and which renders it so noble an ornament of the romantic waters of Killarney.

Andromeda again, with her countless blushing or snow-white flowers, and glossy or powdery evergreen leaves, is like the Arbutus, only the fruit is a dry capsule, opening by valves.

Rhododendron and *Azalea*, on the contrary, have

corollas that spread open at the mouth, with unequal divisions, and with stamens bent towards one side; while *Kalmia* has cup-shaped corollas with ten little niches in which the anthers are nipped up, so that when the flower expands, the filaments are all curved downwards away from the pistil, as if the corolla was unwilling to allow them to touch it; but stir the filament with a pin at the time when the anther is ready to shed its pollen, and in an instant the stamen starts up, and approaches the anther to the stigma.

All these plants are so well known, from being the pride of the American garden, that I have only to name them to recall them to your memory. Considering how very handsome they are, and how innocent is their aspect, you would scarcely suppose that venom lurked beneath their charms; they will, however, serve as an instance of how little you may trust to appearances, even among flowers; for both the Rhododendron, the Kalmia, and the Andromeda, have not only noxious leaves and branches, but their very honey is poison; as has been too fatally experienced by those who have fed of the produce of the hives of Trebizonde.

Extremely different from these is the tribe of *Bindweeds* (Plate XII. 2.), of which the wild *Convolvulus*, at once the pride and pest of our English hedges, and the not less beautiful but more harmless *Ipomœa* of the gardens are the representatives. These plants, like the Heath tribe, are monopetalous, but they have a twining stem, and corollas that are neatly plaited

when they close, like the paper purses that are made
for children. These corollas open and close under
the influence of light or darkness, some opening
only in the day, others only in the night, and in one
case (Ipomœa sensitiva) they are so sensitive, that
they contract beneath the touch like the leaves of
the Mimosa. The calyx of the Bindweed consists
of five sepals, which overlie each other so completely,
that you can seldom perceive more than the two
outermost. The fruit (*fig.* 5.) contains three or four
cells, and a very small number of seeds, the embryo
of which (*fig.* 7.) is doubled up in the most curious
way, just as if there were not room enough within
the seed for it to grow. The roots of many of them
are large and fleshy; they possess powerful medicinal
properties, and are fit for food only in the case of the
Sweet Potatoe (Convolvulus Batatas), which was so
much esteemed before the common Potatoe displaced
it in Europe.

To this tribe also belongs an odd little plant called
Dodder (Cuscuta). Have you never remarked upon
the stems of the Heath, or Nettles, or of the Furze,
clusters of stout reddish cords, which are so twisted
and intertwined that you would take them for a knot
of young snakes, if the colour first, and then their
touch did not undeceive you. If ever you have remarked so strange an appearance you have seen
Dodder, which originally earthborn, soon lays hold
of some neighbouring plant, twists her leafless shoots
around it, fixes them firmly to the branches, quits her
hold of the soil, and thenceforward, as if ashamed of

her humble origin, feeds only upon dews and rain ; till the frost comes, nips her tender frame, and leaves her dead and shrivelled form still clinging to its place, a monument of the punishment of vegetable ambition. This strange plant is of the Bindweed tribe ; but is extremely imperfect ; leaves it has none, except a few stunted scales, and its flowers are little white things collected in close clusters. The fruit consists of membranous capsules, in each of which are two cells and four seeds.

I must now leave you to hunt for Dodder, and to study her singular habits if you can find her, till I have leisure to resume my pen.

EXPLANATION OF PLATE XII.

I. The Heath Tribe.—1. A shoot of *Hispid Heath* (Erica hispida).—2. A flower; *a*, the sepals.—3. The stamens and pistil.—4. and 5. Anthers.—6. A pistil cut open, shewing the arrangement of the ovules within the ovary.—7. The ripe fruit of *Rhododendron*, natural size ; *a*, a central receptacle of seeds, from which the valves separate.— 8. A seed very highly magnified; *a*, the hilum, or scar where the seed separated from the receptacle.—9. The same, cut through perpendicularly, shewing the embryo lying in the midst of albumen.

II. The Bindweed Tribe.—1. A shoot of *Dwarf Convolvulus* (Convolvulus tricolor).—2. The stigma.—3. A stamen.—4. The ovary divided perpendicularly, shewing the manner in which the ovules grow. —5. The ripe fruit.—6. A seed.—7. The same cut through perpendicularly ; *a*, the radicle of the embryo ; *b*, the hilum, or scar where the seed separated from the receptacle.

LETTER XIII.

THE GENTIAN TRIBE—THE OLIVE TRIBE—THE JASMINE TRIBE.

Plate XIII.

IF there is any one tribe of plants in nature, the colours of whose flowers are more intensely vivid, and the foliage neater, and the whole aspect prettier than any other, it is that which comprehends the different species of Centaury, and the beautiful Alpine Gentians with their flowers of azure or yellow. This which is called the *Gentian tribe*, belongs to the Monopetalous division of Dicotyledonous plants, among which it is usually known by the leaves, which are opposite each other on the stem, being *ribbed* (that is having two, or four, or more strong veins parallel with the midrib), and extremely bitter. The flowers are also constructed upon a peculiar plan. The calyx consists of four or five sepals more or less united, the corolla has the plaited appearance that I mentioned in the Bindweeds, with all its divisions equal to each other; there are four or five stamens, and a superior ovary, with two many-seeded cells, and a two-lobed stigma.

All this you will find in the *Gentianella* (Gentiana

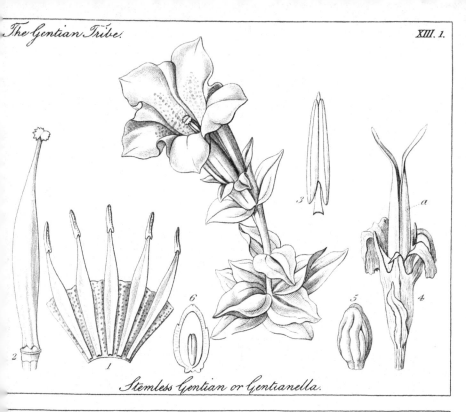

Stemless Gentian or Gentianella.

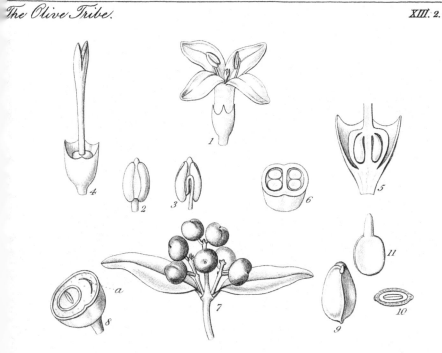

Privet.

acaulis, Plate XIII. 1.), a plant that you cannot fail to procure from the first good garden you
enter. It indicates the presence of the bitter sto-
machic qualities for which the Gentian of the shops
is so much employed, and you may be quite sure
that they exist in any wild species in which a similar
structure is discoverable. Thus we have in our
marshes what is called the *Calathian Violet* (Gentiana
Pneumonanthe), with narrow leaves, and a corolla
greenish externally, but a lovely azure within ; and
on our hills, or sea-cliffs, a branchy dwarf plant with
rose-coloured blossoms (Erythræa Centaurium), called
Centaury ; in both these the same bitterness occurs,
and they both may be used just as well for domestic
bitters as the exotic drug of the shops.

The Gentian tribe is not a very extensive one ; the
principal part of it is met with in the tropical countries of South America; a few species, called *Chironias*,
from the Cape of Good Hope, exist in green-houses ;
another group, of almost uncultivable habits but
great beauty, called *Sabbatias*, is found in North
America ; and the remainder, which are chiefly
Gentians, are found all over the milder and more
alpine parts of Europe and Asia ; in the Swiss Alps,
on Caucasus, and on the Himalayan mountains of
India, they enamel the sward with blossoms of such
intense brilliancy, that the eye can scarcely rest
upon them.

From some unexplained cause the plants of this
tribe are generally so difficult to manage in England,
that with the exception of three or four robust species,

the most skilful gardeners cannot keep them alive :
were it otherwise, there are none which it would be
so easy and so desirable to procure from foreign
countries.

Brief as my remarks upon this tribe have been,
you will find that they are sufficient to enable you to
recognize it. Let me now turn to another.

It is a very unusual circumstance for monopetalous
plants to have only two stamens, unless the corolla is
irregular : that is to say, unless the parts of the corolla
are of unequal size. There are, however, two natural
orders of which the essential character consists in the
presence of two stamens within a regular corolla.
Of these the *Olive tribe* is the more remarkable, and
that which I shall take for illustration. The Olive
itself is so uncommon in England, that it will be more
convenient to select the *Privet* for the plant by which
your notions of the tribe are to be formed.

The *Privet* is a Dicotyledonous shrub, with opposite
leaves. Its calyx is a four-toothed cup : being composed of four sepals united, except just at the tips
(Plate XIII. 2. *fig.* 4.). The corolla consists of four
equal petals, united half-way into a tube (*fig.* 1.),
and joining, before they expand, by their edges only.
The stamens are two, of a very common appearance.
The ovary is superior, and contains two cavities, from
the top of each of which hang two ovules (*fig.* 5. & 6.) ;
it is terminated by rather a thick style, and a two-
lobed stigma To this succeeds a small round black
succulent fruit (*fig.* 7. & 8.), which usually contains
but one seed. This is all that you find in any others of

the Olive tribe, the organization of whose flowers is remarkably uniform.

For example, the *Olive* has a shorter corolla and a hard bony nut in its fruit ; the *Phillyrea*, with its beautiful deep-green leaves, is exactly like the Olive in the structure of its fructification, but its nut is brittle instead of bony ; and the fragrant *Lilac* (Syringa) differs from all these in its longer corollas, and in its fruit being dry, and splitting into two valves.

Simple as is the character of the Olive tribe, and uniform as the genera usually are in their structure, there is one most remarkable exception that I should not omit to notice. The *Ash* (Fraxinus), which you know by its smooth and graceful trunk, and by the airy appearance of its light and elegant foliage, is a plant without any corolla, and yet it belongs to the Olive tribe. It may seem exceedingly strange that a plant which has no corolla should be classed with those which have a perfect monopetalous one, and if such a thing were to happen in an artificial system it would be extremely improper, but in a natural arrangement all *single* characters are subordinate *to the mass of characters*, and, when they do not accord with the usual structure, form exceptions to general rules. Thus, as the Ash agrees with the Olive tribe in every character except the absence of the corolla; that absence is only reckoned an exception to the general fact that the Olive tribe has a monopetalous corolla. It happens that we possess a striking proof, beyond what the fructification affords, that the Ash and Olive are both very nearly related to each other.   It is well known that no tree can be either budded or grafted upon another, unless they are extremely nearly related by natural ties;   the Olive may be grafted upon the Ash, and consequently the inference that is drawn from the construction of the flowers is confirmed by the physical properties of the two plants.

I told you but a short time since that there were two monopetalous natural   orders with   regular flowers, in which there are only two stamens.   The one to which as yet no allusion has been made is the *Jasmine tribe*, to which belong the many fragrant plants which bear that name.   These are known by a most simple character from the Olive tribe; the edges of the divisions of their corolla, instead of being exactly fitted to each other before the flowers expand, overlie each other in the bud, and slide off each other when they unfold.   These differences give rise to two technical expressions which I cannot do better than explain to you on the present occasion.

We call the manner in which the parts of the flower are folded up in the bud the *æstivation;* and we apply the term either to the calyx or corolla, or stamens, or pistils, with some qualifying adjective. When two parts are placed together, edge to edge, so that one does not lie at all upon the other, those parts are said to be *valvate,* and when they do lie upon each other, they are said to be *imbricated,* or tiled, in allusion to the manner in which tiles (called in Latin *imbrices*) are placed upon the roof of a house.   These terms are frequently coupled in speaking of the corolla : of which the Olive and Jasmine tribes afford a striking example. The former is said to have a *valvate æstivation,* the latter an *imbricated æstivation.*

EXPLANATION OF PLATE XIII.

I. THE GENTIAN TRIBE.—A plant of *Gentianella* (Gentiana acaulis).—1. The lower part of a corolla, with the five stamens attached to it.—2. A pistil.—3. An anther.—4. A ripe fruit dividing into two valves *a,* and invested with the withered remains of the calyx and corolla.

II. THE OLIVE TRIBE.—1. A flower of the *Privet* (Ligustrum vulgare).—2. The face of an anther.—3. The back of one.—4. A calyx with its pistil.—5. The same cut through perpendicularly, shewing the ovules hanging in the cells of the ovary.—6. A horizontal section of the same.—7. A cluster of ripe berries.—8. A fruit cut through, with a view of the single seed and the embryo within it; *a* the second cell, which is nearly obliterated by the pressure of the seed upon it.—9. A seed extracted from the pulp.—10. The same cut across.—11. The embryo taken out.

LETTER XlV.

(Plate XIV.)

FROM the Olives of Italy, with their dingy foliage,
and imperishable wood, let us turn to our own innocent
native *Hare-bells*, whose modest beauty amply recom-
penses us for the absence of the gaudy, scented, and
often venomous flowers of more southern climates.
In this plant we find the representative of an extensive natural order, the species of which are scattered
over all Europe and the cooler parts of Asia and
America, dwelling in dells and dingles, by the banks
of rivers, in shady groves, on the sides of mountains,
and even on the summit of the lower Alps, where the
last lingering traces of vegetation struggle with an
atmosphere that neither plant nor animal can well
endure.

We know the *Hare-bell tribe* only in its humblest
state, bedecked with no other ornament than a few
blue or purple nodding flowers ; but in foreign
countries, it acquires a far more striking appearance.
On the mountains of Switzerland, there are species
with corollas of pale yellow, spotted with black ; on

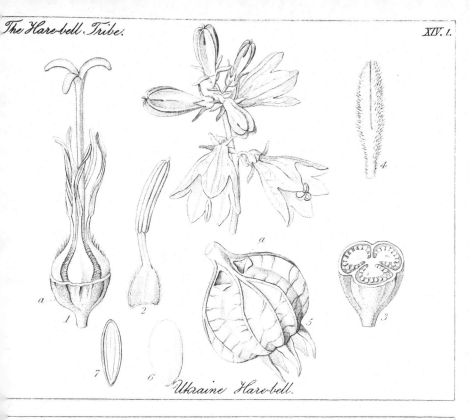

Ukraine Hare-bell.

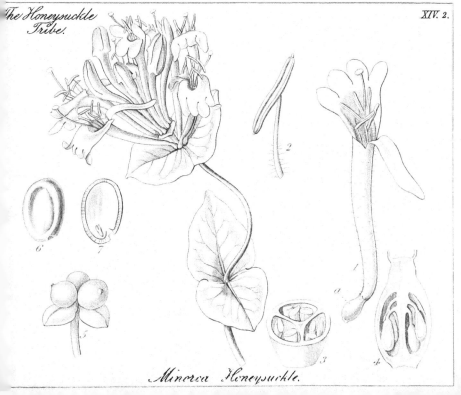

Minorca Honeysuckle.

the Alps of India are others of the deepest purple that can be conceived ; on the rocks of Madeira lives one which was formerly not uncommon in our gardens (Musschia aurea), whose corollas are of a rich golden yellow ; and finally, in the pastures of the Cape of Good Hope are Roellas, the flowers of which are elegantly banded with streaks of violet or rose passing into white.

Let us, however, confine ourselves, in the first instance, to the true *Hare-bell* genus. In every shady lane there grows a diminutive herb, with little grassy leaves, and a few blue bell-shaped nodding flowers ; this is the real *Hare-bell*, which Botanists call the *round-leaved Campanula ;* you will wonder why it is so called, since its leaves are narrow, like those of a grass ; but if you pull it up by its roots, you will then find that the lowest of all the leaves have a roundish outline, from which circumstance it derives its name. You who live in the country will take this species for examination ; but I am obliged to step into a garden for a subject, and I have selected a species found in thickets in the Ukraine, from which it is named Campanula ucranica ; either will answer the purpose equally well.

The calyx of the plant has five deep divisions (Plate XIV. 1.), which spread regularly away from the base of the corolla, and from the top of the ovary. The corolla has very regularly the figure of a bell, except that it is too narrow at the base ; its border is divided into five lobes, which shew that it is made up of five petals, and it is veined in a pretty and peculiar manner. From the base of the corolla, and consequently from the summit of the ovary, spring five stamens (*fig.* 1.), whose filaments are broad, firm, and fringed (*fig.* 2.), curving inwards at the base, and bending over the top of the ovary, as if to guard it from injury; their points touch the style, and keep the anthers parallel, and in contact with it, till they shrivel up and fall back, which happens immediately after the flower unfolds. The style is a taper stiff column, about the length of the corolla, and longer than the stamens. It is covered all over, up to the very tips of the stigma, with stiff hairs (*fig.* 4.) which nature has provided to sweep the pollen out of the cells of the anthers, as the style passes through them in lengthening; if it were not for this simple but effectual contrivance, as the anthers burst as soon as ever the corolla opens, their pollen would drop out of the nodding flowers and be lost before the stigma was expanded and ready to receive the fertilizing influence; the hairs of the style catch the pollen and keep it till insects, wind, or accident brush it down upon the inverted stigmas.

Next let us look at the ovary. This organ is in the *Hare-bell* a case containing three cavities or cells (*fig.* 3.), surrounding a central axis; in each cell there is a large fleshy receptacle, over which is spread a multitude of ovules. After the stigma is fertilized the corolla and the stamens drop off, the sepals harden, enlarge, and collapse, all the parts become browner and thicker, stout ribs appear in the substance of the ovary, which droops still more than the flower itself (*fig.* 5.), and at last a general dryness, hardness, and brownness, announce that the ripening of the fruit is accomplished. But how are the dust-like seeds ever to find their way out of this lidless box, or to penetrate its tough sides? Considering what happens in so many other plants, we should naturally expect that it would take place by a separation of the edges of the three carpels into valves, near their points; but upon looking at the top of the ovary between the sepals, we find that part still tougher than the sides, and without the slightest appearance of an opening. It is by a rending of the thinnest part of the sides of the fruit, in the fork of the three principal ribs (*fig.* 5. *a.*), that these valves are produced, and that nature provides for the escape of the seeds; the rending takes place upon the final drying of the sides of the fruit, when every part becomes stretched so tight, that any weak portion must of necessity give way. As the stretching takes place with uniformity, and as the skin at the forks of the ribs is always more tender than any other part, the opening of the valves will consequently occur with the same invariable certainty as the formation of the seeds.

Do not, however, suppose that this curious contrivance is characteristic of the Hare-bell *tribe:* on the contrary, it is only characteristic of the Hare-bell *genus;* for in other genera, the fruit opens by a separation of the points of the carpels in the usual way; the tension of the sides consequently does not take place, and no lateral openings being necessary, none are ever formed.

Among the commoner genera, allied to the Hare-bell, should be distinguished *Phyteuma*, some very pretty species of which are found on the Alps of Europe, and two, even in hedges in the south of England, although very rarely ; this genus is known from the Hare-bell by its corolla not being bell-shaped; but split into five very long and narrow segments. *Wahlenbergia* also, with the corolla of a Hare-bell, but with the fruit opening at the points, is found plenti-fully in Cornwall and Devonshire, in the shape of a charming little ivy-leaved plant, creeping among the turf, above which it raises its blueish drooping bells (Wahlenbergia hederacea); and finally in cornfields, often enamelling the stubble in harvest time, appears the *hybrid Looking-glass flower* (Specularia hybrida), the corolla of which spreads flat round the stamens, forming little rays with its petals ; its fruit sheds its seeds through three slits in its angles.

The Hare-bell tribe is as harmless as it is beau-tiful ; the roots of some species are eaten under the name of Rampion, the leaves of others are used in Salads, and the bells afford an abundant supply of honey to the bee. The stems and roots abound in a milky juice, which although in this case innoxious, is usually a symptom of poisonous properties, and which, in a neighbouring tribe, indicates the pre-sence of the most fearful venom. As the gardens contain many species of the deleterious group, called the *Lobelia tribe*, I cannot do better than take this opportunity of explaining to you how you may know them.

Imagine a Hare-bell with its corolla split into an

irregular form, and its anthers grown together into a cylinder, through which the stigma projects, and you will have a *Lobelia*, many of the species of which, such as *L. Cardinalis, fulgens* and *splendens*, are the admiration of gardeners on account of their velvety scarlet flowers. Put not your faith in these, for they are all acrid in the most intense degree, and fatal alike to animals and man. They are very common in tropical countries, and are chiefly American ; some of them are herbs like those we see in the gardens, others are bushes, or even small trees. Two of the rarest of British plants are the *burning Lobelia* (L. urens) of Devonshire, and the *water Lobelia* (L. Dortmanna), which inhabits the very bottom of mountainous or northern lakes.

Resembling the Hare-bell tribe in its inferior ovary, but far different in its essential characters, is the tribe which takes its name from the *Honeysuckle*. To understand the structure of this, you cannot do better than study the Honeysuckle itself (Plate XIV. 2.). The leaves of that plant are placed opposite each other with great uniformity ; the uppermost even grow together at their base ; in no case do you find even a trace of stipules, or any thing like them ; a material point which I must beg you to remark.

The flowers have a roundish green inferior ovary (*fig.* 1. *a.*), terminated by a very minute five-toothed calyx, and containing three cells (*fig.* 3.), in each of which hang two or three ovules. The corolla is a tube, the end of which divides into two lips (*fig.* 1.), one of which is narrow and undivided, the other cut

into four rounded lobes: in reality, it is composed of
five petals, one of which is separate for about a quarter
of its length, and constitutes the undivided lip, while
the other four are united nearly to their very points,
and form the upper lip.   Five stamens arise from the
tube of the corolla; the style is long and thread-
shaped, and ends in a pin-headed stigma.   The fruit
is a succulent berry, containing one or two bony seeds
(*fig.* 6.).

Such is the Honeysuckle, the essence of whose
character, consisting in having an inferior, many-
celled, few-seeded ovary, and monopetalous flowers,
is found in Symphoria, one of whose species bears
balls of snow-white fruit, whence it has gained the
name of *Snow-berry*; and in the *St. Peter's Wort*
(Diervilla), the fruit of which is dry and tapering.
It is here also that is stationed the interesting *Linnæa
borealis*, with its delicate rosy bells, and creeping
stems ; by which it is said that the humble and
neglected fate and early maturity of the great Swedish
naturalist, whose name it bears, were typified, at the
time it received its modern title.

Do not imagine that because the Honeysuckle
twines, and Linnæa trails, all the tribes are twiners
or trailers.   On the contrary, if you are acquainted
with either the Snow-berry or the St. Peter's Wort,
already mentioned, or with the *Tartarian and Fly
Honeysuckles* of the gardens, you will already be
aware that many species are upright branching
bushes.   This is more particularly the case with
some other genera.   The *Elder*, of which I long

since gave you a description (page 35), warning you
not to mistake it for an Umbelliferous plant, belongs
to the Honeysuckle tribe; and so do the *Wayfaring
tree* (Viburnum Lantana), and the *Guelder Rose*
(Viburnum Opulus), both of which are to be met with
in every shrubbery. At first sight, these plants seem
to be unlike the Honeysuckle, but study their structure carefully, remembering what it is that I have
told you is essential to the tribe, and you cannot
mistake their affinity. In fact, if they were twining
you would never have doubted it.

In North America there grows a plant of this tribe
with broad leaves, clusters of flowers sitting close in
their bosom, and yellow berries, called Triosteum
perfoliatum, the seeds of which have proved the best
of all substitutes for Coffee. Knowing nothing of the
latter plant, you cannot have suspected that it was in
any way allied to the Honeysuckles; but the fact I
have now mentioned may excite a suspicion that it
may be so, as in reality it is.

Coffee (Coffea Arabica), the infusion of whose
seeds forms the beverage which is probably the most
universally grateful of all that the luxury of man
has prepared, belongs to a very extensive natural
order, almost confined to the warmer parts of the
world, comprehending the meanest weeds and the
most noble flowering trees, obscure herbs with blossoms that it almost requires a microscope to detect,
and bushes whose scarlet corollas are many inches
long; and producing drugs invaluable to man for
their important medicinal properties. Ipecacuanha,

N

Coffee, and various kinds of fever barks, especially
that of Peru, are among its useful products.   Now, if
you gave the Honeysuckle tribe well defined stipules
at the base of the leaves, you would convert them into
plants of the *Coffee tribe ;* for, notwithstanding many
other differences in particular instances, the two
natural orders, viewed in a general manner, can
hardly be said to be absolutely distinguishable by
any other character.   I would, therefore, recommend
you to take this as the true distinction, and not
to trouble yourself about further differences, unless
you intend to study Botany minutely.   Coffee itself
consists of the seeds of the plant, divested of their
skin, and of a dark purple fleshy rind that enveloped
them.   They are formed almost entirely of albumen,
in the base of which a very small embryo is placed.
A seed of any common Honeysuckle *(fig.* 6. & 7.
Plate XIV. 2.) will shew you this; for, in all that
regards the seed, the Coffee tribe and the Honeysuckles agree.   I have said that Triosteum has proved
the best of all substitutes for Coffee ; a circumstance
that will not now surprise you ; but it is probable
that other plants of either the Honeysuckle or Coffee
tribes, would answer the purpose equally well, provided their seeds are large enough, and their albumen
of a hard horny texture.

EXPLANATION OF PLATE XIV.

I. The Hare-bell Tribe.—A few flowers of the *Ukraine Hare-bell* (Campanula ucranica).—1. The ovary *a*, with the stamens and style, in the state in which they are found after the corolla has expanded.—2. A stamen with its broad thick base.—3. A horizontal section of the ovary, shewing the three cells and many seeds.—4. A view of the hairy style and stigma, before the lobes of the latter separate.—5. A ripe fruit; *a*, the holes through which the seeds escape.—6. A seed.—7. The same cut perpendicularly, shewing the embryo.

II. The Honeysuckle Tribe.—A piece of the *Minorca Honey-suckle* (Caprifolium implexum).—1. A flower, with the inferior ovary.—2. An anther, with the upper end of the filament.—3. A horizontal section of an ovary, with the three cells, and the ovules cut through as they hung in them.—4. A perpendicular section of an ovary, shewing how the ovules hang from the top of the cavity.—5. A little cluster of fruit.—6. A seed.—7. The same cut through perpendicularly, shewing the embryo.

LETTER XV.

THE BORAGE TRIBE—THE NIGHTSHADE TRIBE— THE PRIMROSE TRIBE.

Plate XV.

I HOPE that you will have found the distinguishing characters of all the Monopetalous orders we have examined up to this time sufficiently clear and definite to be understood and remembered ; and that as I have been proceeding, you have been analysing their distinctions after the plan of which some instances have been already given you.   For I am convinced by long experience that this is the only sure way of fixing such matters in the memory.   As it is my intention, after we shall have gone through the whole of them, to analyse for you such of the Monopetalous orders as I may select for illustration, it is unnecessary for me to dwell as yet upon their mutual distinctions. Let us, on the contrary, proceed, for the present, steadily in our examination of other natural orders.

Of all the groupes into which Botanists have divided the Vegetable Kingdom, there is none which combines uniformity of general appearance with similarity in structure in a greater degree than those rough-leaved plants, which Linnæus used, on

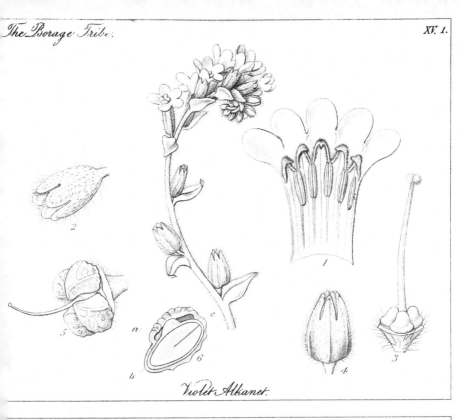

Violet Alkanet.

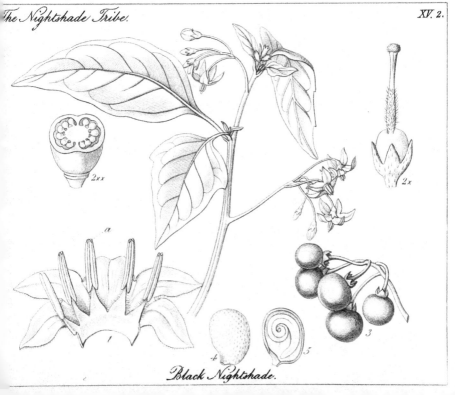

Black Nightshade.

that account, to call Asperifoliæ, and which we know
by the name of the *Borage tribe*. Leaves covered
with rigid hairs, a regular monopetalous corolla, and
a four-lobed ovary, which changes into four seed
like grains, form the peculiar character of this natural order, of which the *Violet Alkanet* (Anchusa italica) is a common and good illustration. In Botanic
gardens, or in collections of biennial plants, this is
an extremely common plant, so that I anticipate no
difficulty in your procuring it. Should you, however, be unable to procure it when this letter reaches
you, take a shoot of *Forget-me-not*, by which you
may also follow me. My reason for selecting the
Alkanet, is the large size of all its parts of fructification.

The leaves of this plant have a fleshy texture and
a mucilaginous pulp ; but their skin is covered with
hairs so stiff and sharp that they will prick the
fingers if drawn over them against the hairs. A
microscope will shew you that this is owing not so
much to the stiffness of the hairs themselves, as to a
hard stony base from which they arise, and which
if your magnifying power is strong enough, will be
found to consist of a cluster of very hard cells of
cellular tissue. When the leaves are young, this
hardness is less remarkable ; but as they grow old,
it becomes very conspicuous. The leaves are placed
alternately on the stem, and the latter is round—two
facts which I will beg you to recollect.

The flowers are arranged in a singular manner.
The stalk which bears them is coiled up (Plate XV. 1.)

at the point, so that the youngest flowers are quite hidden by its folds ; but it gradually uncoils as the flowers expand, till at last it becomes nearly straight. In consequence of this singular arrangement, the flowers are all forced towards one side, and when they are expanded, look as if they actually grew from one side only ; this, however, is not the fact—they are only turned towards one side. I think I have already explained to you that we call the arrangement of the flowers upon their common stalk the *inflorescence;* and that different adjectives are added to this word to explain its nature. Thus, in Umbelliferous plants, the inflorescence was umbelled ; in the Borage tribe it is what is called *gyrate;* a fine word expressive of being coiled. Now this gyrate inflorescence will of itself enable you to recognise the Borage tribe, and the families most immediately allied to it, without recourse to any thing further.

The calyx, which is covered with hairs, like the leaves, consists of five sepals joined to each other more than half way, so as to form a tube *(fig. 2.)*. The corolla has its border divided into five lobes, opposite which at their base, are five hairy convex scales, which converge and close over the mouth of the tube, so as effectually to prevent any intruder from entering it *(fig. 1.)*. From the side of the tube of the corolla, below the scales, spring five stamens, which sit close upon its surface, without any visible filament. The ovary is divided into four deep lobes *(fig. 3.)*, from the middle of which rises a taper style, terminating in a double stigma.

When the fruit is ripe, it is invested with the calyx, which remains green for a long time (*fig.* 4.), only contracting at the point, so as to cover the fruit. Corresponding with the four lobes of the ovary, are four grains, or rather rugged bony nuts (*fig.* 5.), which finally separate from each other, when they look like so many seeds, for which they used to be mistaken. These, along with the gyrate inflorescence, are the great characters of the Borage tribe, as distinguished from all other Monopetalous Natural Orders.

Many of the genera of the Borage tribe are extremely common. The most beautiful of all our wild flowers, *Viper's-Bugloss* (Echium vulgare), is one of them. It is known by its corolla having an unequal and irregular margin, and a sort of bell-shaped figure. The deep red or purple spots and hairs of that plant are very remarkable.

Then there is *Forget-me-not* (Myosotis), whose various species ornament ditches and dry banks with their pretty blue blossoms. It is known from Alkanet only by the scales of the mouth of its corolla being more rounded and shorter.

Borage, too (Borago officinalis), occasionally makes its appearance upon banks, and in waste places; it is conspicuous for its azure flowers, whose corolla is deeply divided into five spreading lobes, which are much longer than the tube.

Besides these, we have *Hound's-tongue* (Cynoglossum), with long grey leaves, and dingy reddish brown flowers, succeeded by broad fruit, covered all over with stiff hooks; *Lungwort* (Pulmonaria), with leaves

spotted with green and white, and flowers of a lovely
blue, shaped like a funnel; the mean-looking *Grom-*
well (Lithospermum), which seems as if conscious of
its worthlessness, by constantly dwelling with weeds
and rubbish; whose very fruit is so like a stone, that it
derives its Botanical name (Lithospermum signifies
literally stone-seed) from that circumstance; and
finally *Comfrey* (Symphytum), with tall coarse stems,
and tubular flowers, the scales of whose mouth seem
as if made expressly to teach us what the real nature
of those singular parts are in other genera. In
the Comfrey, the scales are so exactly like the
filaments, that if you cut off the anthers, you cannot tell one from the other; and consequently, as
all other circumstances confirm the opinion of
their being abortive stamens, the scales are so considered.

Not one of the Borage tribe is otherwise than harmless; the young shoots of Comfrey have even been
eaten as Asparagus, but they have too little taste to
be worth cultivation; and Borage itself was once an
ingredient in a favourite beverage of our forefathers,
called a cool tankard. It is however for their dyeing
properties that they are really valuable. The dye
called Alkanet, or others of a similar quality, is furnished by the roots of Anchusa tinctoria, Lithospermum tinctorium, Onosma echioides, and several other
species.

From this harmless natural order, let us turn to
one, the properties of which are too often dangerous.
Henbane, Nightshade, and Tobacco, the narcotic

Thorn-apple, with the half fabulous Mandrake, whose roots were said to shriek as they were torn from the earth to give effect to magical incantations, form, with a number of other plants, a large natural order, the prevailing quality of which is to be poisonous. Many of them are common wild plants, and none more so than the species called *Black Nightshade* (Solanum nigrum), which is sure to spring up wherever a spot of ground is neglected, and suffered to become waste. It is to this plant I shall trust for explaining the general characters of the *Nightshade tribe.*

Black Nightshade is a plant with broadly lance-shaped leaves, slightly toothed at the edge, and seated alternately upon the stem (Plate XV. 2.). Its flowers consist of a short five-toothed calyx, of a monopetalous corolla, with five equal divisions (*fig.* 1.), of five equal stamens, and of an ovary (*fig.* 2**.) with two cells, in each of which is a number of ovules. The style of the ovary is thick and shaggy at the bottom, and terminated by a thickened undivided stigma. The fruit is a small black berry, containing two cells, and a number of yellowish seeds, whose skin is covered closely with little pits (*fig.* 4.); in the inside is an embryo, which is coiled up upon itself in the middle of a quantity of fleshy albumen (*fig.* 5.). Of these characters, the most essential ones are the *superior ovary with two cells, the regular flower, and the alternate leaves.* The last point distinguishes the Nightshade tribe from the Gentians, and you will find, by and bye, that the two former separate it from other orders you have still to examine.

The genus Solanum is known in its tribe by the anthers opening by two holes or pores at their points (*fig.* 1. *a.*); besides Black Nightshade, it contains the *Bitter-sweet* (S. Dulcamara), whose red and tempting berries present a dangerous decoy to children; the *Love Apple* (S. Lycopersicum), or *Tomato*, the pulp of which is so much esteemed in sauces; the *Egg-plant*, or *Aubergine* (S. Melongena), whose fruit, when fried in slices, forms a delicacy in French cookery; and above all, the *Potatoe* (S. tuberosum). Here, you will imagine, is a singular assortment of eatable and poisonous plants in the same genus; but in truth, the fruit of these is in all cases deleterious till it is cooked; Tomatoes are stewed, Egg-plants are washed and fried before they are eaten, and it is not to be doubted that they would all prove injurious, if used in a raw state. The fruit of the Potatoe is notoriously unwholesome; and if its roots are not so, that circumstance is to be ascribed in part to their being cooked, and in part to their being composed almost entirely of a substance like flour, which in no plants is poisonous, if it can be separated either by heat or by washing, from the watery or pulpy matter it may lie among.

Deadly Nightshade (Atropa Belladonna) belongs to a genus resembling the last in its berries, but having a bell-shaped flower, and anthers which open by slits in the usual way. The fruit of this plant is the most venomous of all our wild berries; it is of a deep

shining black, and follows a livid brown corolla.
The *Mandrake* is a species of the same genus (Atropa
Mandragora), but it has whitish flowers veined with
purple, and scarcely any stem; it is only found in
the southern parts of Europe.

Henbane (Hyoscyamus niger) lives on commons,
especially where the soil is chalky, near old cities and
upon banks. Its broad pale leaves have a fetid
smell, are irregularly lobed at the edges, and are
covered all over with greasy hairs. The flowers sit
close upon the stem, and have a large dirty-yellow
corolla, veined with brownish purple, which gives
them a peculiar livid appearance.

The *Thorn-apple* (Datura Stramonium), so celebrated for its narcotic properties, is distinguished by
its fruit being dry and covered with stiff spines; and
Tobacco (Nicotiana Tabacum) by its long tubular
corolla and smooth dry fruit bursting into two valves.

To enumerate any greater number of these deleterious plants would be to occupy a larger space than
I have room for; what I have already mentioned
will suffice to make you understand their general
nature; and for the rest you must consult systematic
works. All that I beg you to recollect is, that plants
of the Nightshade tribe, are not only monopetalous,
but have *a superior two-celled ovary, regular 5-lobed
flowers, and alternate leaves.* The fruit of all such
avoid, until it has been ascertained by the experience
of others, that the ycan be eaten with safety.

You will scarcely suspect that those prettiest of
spring flowers, *Primroses, Oxlips,* and *Cowslips,* can be in

any way related to the venomous plants I have just
mentioned; nor do they in fact belong to the same
tribe; but they are so similar in many respects that
I shall have no opportunity more fitting than the
present, to say a word to you about them. Endeared
as they are to us all by some of the sweetest recollec-
tions of infancy, it would almost amount to a crime
to pass them by with neglect. Like the Nightshade
tribe they have regular monopetalous flowers with
five stamens, and a superior ovary; they are sometimes similar in habit, as in the case of the Mandrake,
which resembles a gigantic Primrose with white
flowers marked by .purple veins, and they also possess
slight narcotic properties. They are distinguished
by one circumstance in particular, by which they
may be at all times known among our wild flowers
with certainty,—*their stamens are not placed between
the lobes of the corolla,* as in the Nightshade tribe, *but
are opposite to them,* a very curious and permanent
difference.

This you will instantly discover by the examination
of the *Primrose,* the *Auricula,* or the *Polyanthus.*
The ovary is also constructed on a different plan:
you will find that of the Primrose to contain only one
cell, with the ovules collected in the centre: in the
Nightshade tribe there are two cells; on the outside
of the ovary of the latter, you will discover two furrows on opposite sides of it, indicating that it is con-
stituted by the growing together of a pair of carpels:
on the outside of the ovary of the Primrose are five
furrows, slight indeed, but sufficiently apparent, and

indicating its formation out of five carpels. The peculiarity in the stamens is, however, sufficient without referring to the fruit, except when the corolla has fallen off.

With this character agree the beautiful pigmy Alpine plants called *Aretia* and *Androsace;* here also are referred *Soldanella* with its little bells of blue so prettily notched on the border, and Cyclamen or *Sow-bread,* whose fruit is forced, by the rigid coiling up of the flower-stalk, down upon the earth, where it lies concealed by the broad ivy-like leaves. Here too are arranged the *Pimpernel* (Anagallis), one of whose species is called the *Poor Man's Weather-glass,* because it is found in every piece of waste ground, and will only open its tiny brick-red flowers in fine weather, closing them at the approach of rain; and *Loosestrife* (Lysimachia), whose creeping stems, little yellowish-green leaves, and brilliant yellow flowers are the brightest ornament of the moss and short herbage that springs up in woods and shady places.

Interesting as are the British species of this natural order, they are far inferior in beauty to their relations who live on the mountains of other countries : for the Primrose tribe most frequently prefers Alpine stations to all others. It is in the higher regions of the mountains of Switzerland and Germany, on the Pyrenees, and upon those stupendous ridges, from which the traveller beholds the vast plains of India stretching at his feet in a boundless panorama, that the Primrose tribe acquires its greatest beauty. Living unharmed beneath a bed of snow during the

cold weather, where it is protected alike from light
and from drying winds, as soon as the snow is melted
it springs forth bedecked with the gayest tints imagi-
nable; yellow, and white, and purple, violet, lilac,
and sky blue are the usual colours of its flowers;
while its leaves, nursed by the food descending from
a thousand rills of the purest water, and expanded
beneath an ever genial and cloudless sky, acquire a
green which no gem can excel in depth or brightness. It is in those regions only that the Primrose
tribe can be studied to the greatest advantage.

EXPLANATION OF PLATE XV.

I. THE BORAGE TRIBE.—A portion of the inflorescence of *Violet
Alkanet* (Anchusa italica), shewing its gyrate disposition.—1. A corolla
opened, exhibiting the stamens and the scales which close its throat.—
2. A calyx.—3. The four-lobed ovary, style, and stigma.—4. The
calyx of the fruit.—5. The four-lobed fruit with the withered style still
remaining.—6. One of the seed-like lobes of the fruit divided perpendicularly; *a*, the part which separated from the receptacle; *b*, the
organic apex; *c*, the organic base; these are inverted in the drawing so
that the section may correspond with the position of the lobes in fig. 5.

II. THE NIGHTSHADE TRIBE.—A twig of *Black Nightshade*
(Solanum nigrum).—1. A corolla laid open; *a*, the holes through which
the pollen is discharged by the anthers.—2*. The pistil and calyx.—
2**. A horizontal section of the ovary, exhibiting the numerous seeds
lying in the two cells.—3. A cluster of ripe fruit.—4. A seed.—5. The
same divided perpendicularly, shewing the manner in which the embryo
is coiled up.

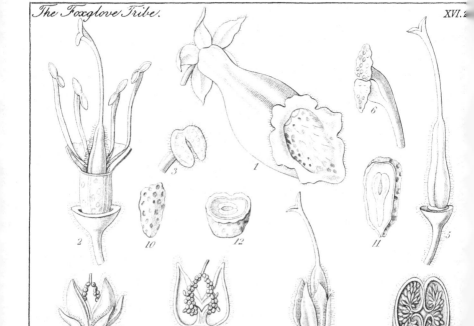

Black Horehound.

Purple Foxglove.

LETTER XVI.

THE MINT TRIBE—THE FOXGLOVE TRIBE.

Plate XVI.

ONE of the easiest of all natural orders to recognise is the Borage tribe, which formed the subject of part of my last letter. Its coiled or gyrate inflorescence and four-lobed ovary are so peculiar, as never to escape the observation of the most careless inquirer. There is only one way in which a mistake may be made by a beginner, and that I shall now teach you how to avoid.

Of all the weeds of common occurrence, that which is perhaps the most universally distributed, and which appears by the strength of its constitution the best adapted to flourish in situations where little else can grow, is *Black Horehound* (Ballota nigra). Even by the sides of the high road, in the midst of the hottest and dryest weather, when it becomes literally cased in a thick coating of dust, that plant flowers, regardless of the elements and their effects. You may know it by its dull dusty disagreeable smell, its roundish indented wrinkled dull grey leaves, placed opposite each other on a square stem, and its whorled clusters of dull purple flowers. The ovary is split

into four lobes ; its fruit consists of four black grains,
which are the divisions of the ovary in a hardened
state, and in fact it is so very like the fruit of a plant
of the Borage tribe, that I am quite sure when you
look at its monopetalous corolla, you will believe that
it belongs to that natural order.

But pursue your examination of it a little further,
and see whether its other characters are also in accordance with those of the Borage tribe.   Its leaves
for example—are they covered with stiff hairs ? no—
are they placed alternately on the stem ? no—have
they the insipid taste, &c. of the Borages ? no—are
the flowers arranged in a coiled or gyrate inflorescence ? is the corolla regular ? are there five stamens ?
to all these questions the answer still is no, no, no.
Then Black Horehound does *not* belong to the Borage tribe.

Look then a little more exactly into the structure
of the flowers ; for there can now be no doubt that it
is a part of some natural order you have not yet examined.   Its calyx (Plate XVI. 1. *fig.* 1.) is a tube
with five sharp-pointed teeth (*a.*), and is consequently
formed of five sepals.   The corolla is hairy and tubular at the base (*fig.* 1.), and divided at the top into
two unequal parts called *lips;* of these lips the upper
is narrow and concave (*b.*), the lower is flat, and divided into three lobes(*c.*), of which that in the middle
is much larger than the two side ones ; this corolla
is therefore *very irregular*.   To see the stamens distinctly, the best plan is to slit open the corolla (*fig.*
2.) ; you will then find that there are four of them,

two being shorter than the others; their anthers consist of two lobes which diverge very much, and are only connected just at the tips (*fig.* 3.). The structure of the ovary and fruit (*figs.* 4. & 6.) is like that of the Borage tribe; but the style is uniformly forked at its upper end, and has a very minute stigma on each point of the fork (*fig.* 5.).

Plants thus constructed, and there is a considerable number of them, form what Botanists term *Labiatæ*, and which may be called in English *the Mint tribe.* They are known from all, except the Borages, by their four-lobed ovary; and from the Borages they are distinguished by their opposite leaves, square stems, irregular flowers, and several other characters; especially by a total difference in their sensible properties. While all the Borages are insipid and scentless, the Mint tribe consists of aromatic herbs, whose leaves and flowers are both impregnated with a volatile matter which is continually exhaling, and which becomes exceedingly perceptible when the parts are rubbed. For instance, *Lavender*, *Thyme*, and *Rosemary*, *Mint*, *Basil*, *Sage*, *Marjoram*, and *Clary*, the common aromatic herbs of the Kitchen Garden, are all relatives of each other, and belong to this natural order.

Of our wild flowers the most remarkable genera, besides those above mentioned, are the *Dead Nettles* (Lamium), with their strong-smelling leaves and flowers of purple or white; *Bugles* (Ajuga), with blue flowers and creeping stems; *Ground Ivy* (Glechoma), that crawls over the bottoms of dry ditches and up the sides of banks among the grass; *Woundwort* (Stachys), which owes its name to the blood-red stains upon its corolla; *Self-heal* (Prunella), peeping up among the sward of commons and old pastures; and *Gipsy-wort* (Lycopus), with its minute pale rosy flowers, which inhabits the banks of rivers and lakes, and yields a deep brown stain when its leaves are bruised in water.

All these and a thousand more agree in having perfectly harmless properties, and are for the most part aromatic; their favourite places of resort are hedges, woods, and shady lanes; they spring up on the sloping face of chalky downs, and enamel the meadows of subalpine regions; even the scorching sun of Syrian deserts they can endure, but they are unable to support the most intense cold in which vegetation can exist. In Melville Island, for instance, situated in the Arctic Ocean, none of the tribe were found by Captain Parry's officers, although Saxifrages, little Potentillas, and many other pretty flowers appear in the summer through the thin coating of soil which conceals the everlasting ice of that desolate region.

Far removed from these in their qualities, for many are poisonous, and all suspicious, although extremely similar in most points of organization, are the beautiful plants which form the *Foxglove tribe* (Plate XVI. 2.). They are known by their fruit not consisting of four seed-like lobes, but being a hollow case, or capsule (*fig.* 7.), containing two cells (*fig.* 9.), and a great number of seeds. In the irregu-

larity of the corolla and stamens, they exactly agree
with the last, only it is carried much farther in the
Foxglove tribe than in the Mint tribe; for the petals are sometimes so irregularly combined, that one
can scarcely make out their real number and nature
by any of the ordinary tests; as in the charming
genus *Calceolaria*, the lower lip of whose corolla is
inflated like the foot of a clumsy slipper, and in the
still more remarkable Chilian *Schizanthus*, whose
corolla is cut and slashed as if it had been clipped
with a pair of shears.

No better example of this order can be selected
than the common *Foxglove* (Digitalis), which is so
striking an ornament of many parts of England. Its
corolla (*fig.* 1.) is a large inflated body, with its throat
spotted with rich purple, and its border divided ob-
liquely into five very short lobes, of which the two
upper are the smaller; its four stamens are of unequal length (*fig.* 2.) ; and its style is divided into
two lobes (*fig.* 6.) at the upper end. A number of
long glandular hairs cover the ovary (*fig.* 5.), which
contains two cells, and a great quantity of ovules.

This will shew you what the usual character is of
the Foxglove tribe ; and you will find that all the
other genera referred to it in books agree with it
essentially, although they differ in subordinate points.
It is chiefly in the form of the corolla, in the number
of the stamens, in the consistence of the rind of the
fruit, its form, and the number of seeds it contains,
and in the manner in which the sepals are combined,
that these differences consist.

Thus *Figwort* (Scrophularia) has a globular corolla, with a large upper, and a very small lower lip; *Speedwell* (Veronica), with its spikes of blue, has only two stamens; *Snapdragon* (Antirrhinum) has the throat of the lower lip so prominent as to press against the upper; *Monkey flowers* (Mimulus) have the angles of the calyx winged; and the woolly *Mulleins* (Verbascum), a corolla with scarcely any tube. No genus is more remarkable than Pentstemon, of which so many fine American species adorn our gardens; in that plant there are actually five stamens, four of which are of two different lengths, as is usual, while the fifth is long and slender, very hairy at the point, and projects into the very mouth of the corolla; it is, however, notwithstanding its size, imperfect: for it bears no anther. This plant is interesting, as shewing that in such irregular flowers as those of the Mint and Foxglove tribes, there is a tendency to become regular, which is sometimes very strongly manifested. To render a Foxglove regular, it should not only have the divisions of its calyx and corolla of the same size, and of the same number; but the stamens should agree in number with those ; that is to say, as there are five sepals and five lobes to the corolla, there should also be five stamens, instead of four, to constitute perfect regularity. It is considered that in those plants in which there are only four stamens, the fifth is abortive or undeveloped, and that when two stamens only appear, the three others are abortive. This will explain to you why a fifth stamen appears in Pentstemon ; and also the nature of a little scale which is very often found at the back of the corolla in these two-lipped corollas ; the scale is a stamen in a rudimentary state.

If you next compare the structure of the Foxglove tribe with that of the Nightshade, you will remark that the resemblance between them is quite as great as between the Foxglove and the Mint tribe ; only in a different way. They both have monopetalous flowers, five divisions of the calyx and corolla, and an ovary with two cells; but in the Nightshade tribe the stamens uniformly correspond in number with the lobes of the calyx and corolla ; while in the Foxglove tribe they are uniformly fewer ; in a word, the flowers of the former are symmetrical and regular, of the latter, unsymmetrical and irregular, and this is the great distinction between them.

With this I must abandon the explanation of the Monopetalous Dicotyledonous natural orders, which have more than one carpel in each flower ; my next letter will be confined to those, the carpel of which is absolutely solitary and simple.

EXPLANATION OF PLATE XVI.

I. THE MINT TRIBE.—1. A flower of *Black Horehound* (Ballota nigra) ; *a* a lobe of the calyx ; *b*, the upper lip of the corolla ; *c*, the lower lip.—2. A corolla split open to shew the position of the stamens. —3. The top of a filament, with its anther.—4. A pistil; *a*, a fleshy disk, out of which springs the four-lobed ovary.—5. The two lobes of the style.—6. A ripe fruit before the lobes separate.—7. One of the lobes apart.— 8. A fruit cut through horizontally, shewing the four embryos at *a.*

II. THE FOXGLOVE TRIBE.—1. A flower of *Purple Foxglove* (Digitalis purpurea).—2. The stamens projecting from the base of the corolla.—3. An anther.—4. The pistil, after the corolla has dropped off.—5. The same without its calyx.—6. The top of the style and the stigma.—7. A ripe fruit, burst into two valves, and leaving the receptacles of the seeds in the middle.—8. A perpendicular section of the same, with the receptacles left.—9. A horizontal section of a half-grown fruit, shewing the precise shape of the receptacles of the seeds.—10. A seed.—11. A perpendicular, and 12, a horizontal section of the same, exhibiting the embryo lying in the midst of albumen. (N. B. These details are after a drawing by Mr. Francis Bauer).

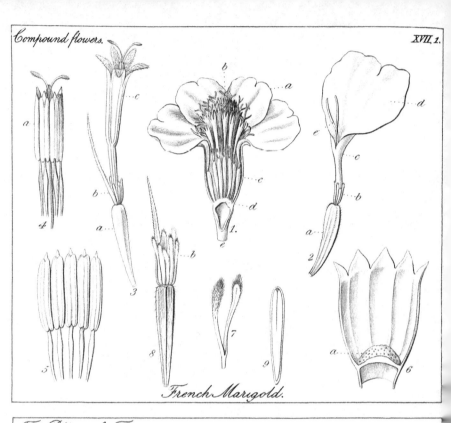

French Marigold.

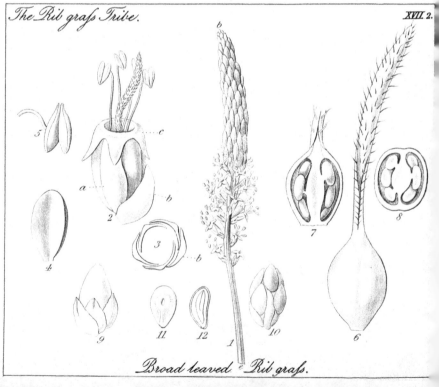

Broad leaved Rib grass.

LETTER XVII.

COMPOUND FLOWERS—THE RIBGRASS TRIBE.

(Plate XVII.)

BEFORE I come to my promised analysis of the orders of Monopetalous plants, I must beg you to give your attention to two more tribes, in which there are many singular points of structure, and which in particular differ from all that have gone before in having only one carpel in each flower; a simple and obvious difference, which you will be sure to remember.

" Take," says Rousseau, whom I shall here follow very closely, " Take one of those little flowers which cover all the pastures, and which every body knows by the name of *Daisy*. Look at it well; for I am sure you would have never guessed from its appearance, that this flower, which is so small and delicate, is really composed of between two and three hundred other flowers, all of them perfect, that is, each of them having its corolla, stamens, pistil, and fruit; in a word, as perfect in its species as a flower of the Hyacinth or Lily. Every one of those leaves, which are white above and red underneath, and form a kind of crown round the flower, appearing to be nothing more than little petals, are in reality so many true flowers; and every one of those tiny yellow things also, which you see in the centre, and which at first you have perhaps taken for nothing but stamens, are real flowers. If your fingers were already exercised in botanical dissections, and you were armed with a good glass, and plenty of patience, I might convince you of the truth of this; but at present you must begin, if you please, by believing me on my word, for fear of fatiguing your attention upon atoms. However, to put you at least in the way, pull out one of the white leaves from the flower; you will think at first that it is flat from one end to the other; but look carefully at the end by which it was fastened to the flower, and you will see that this end is not flat, but round and hollow, in form of a tube, and that a little thread ending in two horns, issues from the tube; this thread is the forked style of the flower, which, as you now see, is flat only at top.

" Next, look at those little yellow things in the middle of the flower, and which, as I have told you, are all so many flowers; if the flower is sufficiently advanced, you will see some of them open in the middle, and even cut into several parts.

" These are monopetalous corollas, which expand, and a glass will easily discover in them the pistil, and even the anthers with which it is surrounded. Commonly the yellow florets towards the centre are still rounded and closed. These, however, are flowers like the others, but not yet open; for they expand successively from the edge inwards. This is enough to shew you by the eye, the possibility that all these small affairs, both white and yellow, may be so many distinct flowers; and this is a constant fact. You perceive, nevertheless, that all these little flowers are pressed, and inclosed in a calyx, which is common to them all, and which is that of the Daisy. In considering then the whole Daisy as one flower, we give it a very significant name, when we call it a *compound flower*. Now, there are many genera and species of flowers formed, like the Daisy, of an assemblage of other smaller flowers, contained in a common calyx. This is what constitutes the sixth tribe, of which I proposed to treat, namely, that of the *Compound Flowers*."

Thus, in his admirable gossiping manner does Rousseau set about explaining the structure of a compound flower; I shall not continue to follow him, but take my own way of illustrating the subject further, stopping, in the first place, to notice two inaccuracies into which he has fallen, in common with all writers of the same day. The cluster of little flowers or *florets*, which he calls the flower of the Daisy, although qualified by the addition of compound, leads unnecessarily to a confusion of ideas, and is much better designated by the term flower-head (or head of flowers), which it really is; the other error is that of calling the little leaves that surround the florets a calyx; he should have said they were an involucre. The word calyx strictly belongs to a single flower, and not to a collection of flowers; while involucre is precisely the term that expresses an assemblage of little leaves or bracts round a number of flowers. You have already had an excellent instance of it in Umbelliferous plants, and this is another.

Rousseau took the Daisy to explain the structure of Compound flowers, because it is so very common a plant; but it has the defect, as a means of illustration, that its parts are so very small, as to be distinguished with difficulty, and that it does not comprehend so many points of structure as some others. For these reasons, I will recommend you to take a plant equally common in gardens in the autumn, the *French Marigold* (Tagetes patula. Plate XVII. 1.), an old-fashioned, but pretty flower, which is not liable to the same objections.

Its flower-head *(fig.* 1.) is surrounded externally by an olive green cup, formed of several bracts which have grown together at the edge (*fig.* 6.); this cup is the involucre. Next the involucre are placed several florets *(figs.* 1. & 2.), whose corolla is a broad yellow blade, rounded at the end, and striped with wide streaks of chocolate brown (*fig.* 2. *d.*); it is all turned one way, spreading away from the flower-head, and only tubular at the bottom; Rousseau says, such corollas look as if they were gnawed off on one side. Technically, they are named *ligulate,* which signifies strap-shaped, because in the greater part of Compound flowers they are long and narrow; they are also said to form the RAY of the flower-head. At the base of the tube of the corolla you will find a few little narrow hairy scales (*fig.* 2. *b.*), which stand on the top of the ovary, in the place of the calyx. Botanists choose to call them the PAPPUS; although in reality they are the calyx, which is only stunted and starved in consequence of its being developed amidst the constant pressure of the florets against each other; the pappus is often altogether absent, as in the Daisy for instance; but it sometimes forms a beautiful plume of feathers, which catches the wind and enables the seed to soar into the air, and to scatter itself to a distance. The delicate feathery balls of the Dandelion, which children amuse themselves with blowing away into the air, are the fruit of that plant crowned by the pappus. Below the pappus is the ovary (*fig. 2. a.*), containing one single ovule; it terminates in a slender style, which passes through the tube of the corolla, and forks at the top into two stigmas (*fig. 2. e.*). In time the ovary becomes a dry hairy fruit (*fig. 8.*), crowned with the pappus, and containing one single seed (*fig. 9.*). Such are the *florets of the ray.*

The middle of the flower-head (*fig. 1. b.*), included within the ray, is called the DISK; it consists of florets constructed very differently from those of the ray. To examine them conveniently you should pull one of them out (*fig. 3.*). In the ovary you will find no difference worth naming; the pappus is also like that of the ray, only it is more perfect, and one of its scales is a sort of stiff bristle (*fig. 3. b.*). The corolla is of quite another kind; it is tubular from the bottom to the top; towards the top it widens, and at last separates into five little divisions which are covered all over with hairs in the inside; this

kind of floret is called *tubular*.　The stigmas are two
(*fig.* 7.), and project beyond the mouth of a little
hollow cylinder, which is found at the orifice of all
the tubular florets of the disk (*fig.* 4. *a.*).　At first
sight you may be at a loss to determine what the
cylinder is : but if you use a magnifying glass, you
will presently discover that it is formed of five anthers,
which grow together by their edges, in the same
manner as petals grow by theirs, when they form a
monopetalous corolla.　It is easy to slit this cylinder (*fig.* 5.), and then you will see that each anther
has its filament, and two lobes containing the pollen.

The broad flat part, out of which the florets grow
(*fig.* 6. *a.*), is called the RECEPTACLE; it is sometimes covered with scales, or hairs, or is even pitted
with hexagonal depressions, which look like the cells
of honeycomb.　Can you guess to what, in other
plants, this receptacle is analogous ? for you may be
sure it is only some very common part masqued and
disguised, to the eyes of an ordinary observer.　To
understand it, you should first compare the flower-
head of a compound flower, with an umbel (See
Letter II.).　In the latter the flowers are all on long
stalks, that proceed from a common point, which
is the termination of the stem or branch; their point
of origin is wider than the stem itself, because of the
number of stalks, for whose bases room has to be
found ; the stalks, however, are very slender, and
do not occupy much room.　But suppose the flowers
of an umbel had no stalks, but were seated close upon
the stem, as sometimes happens; it is obvious that as

they are much stouter than their stalks, the stem
would, in such a case, have to be expanded much
more in order to receive them ; in fact it would
become a receptacle such as you find in a Daisy.
The receptacle of a compound flower is therefore an
expanded part of a stem.

You may also look at its structure in another way.
Take a spike of Ribgrass, an exceedingly common
weed, about which I shall say something at the end
of this letter ; place it by the side of a flower-head of
French Marigold ; let the letter *e* in both cases represent the base of the inflorescence, and *b* the top
of the spike of Ribgrass (Plate XVII. 2. *fig.* 1.),
and the centre of the disk of the French Marigold
(Plate XVII. 1. *fig.* 1.). Suppose the spike of
Ribgrass to be very much shortened, the number of
flowers upon it remaining the same ; the distance
from *b* to *e* will be proportionably diminished, and the
flowers will be much more crowded. Let this shortening be carried still further, the number of flowers
still remaining the same, and it is obvious that in
order to make room for the flowers, side by side, the
stem must expand horizontally ; a receptacle will
then be produced, and a little reflection will shew you
that the letter *b* will then indicate precisely the same
parts in both plants : the centre of the French Marigold being the same as the point of the spike of Ribgrass. The receptacle of a compound flower is therefore both a contracted and expanded stem.

With regard to the involucre, Rousseau has well observed, that it has generally the property of opening

when the florets expand, of closing when the corollas fall off, in order to confine the young fruit; and lastly, of opening again and turning quite back to give more room to the fruit, which increases in size as it grows ripe. This is particularly remarkable in the *Dandelion.*

Let us now pass from these considerations to a view of the sections into which Compound Flowers are naturally divided.

If you gather a head of *Dandelion* (Leontodon Taraxacum), you will find that both the ray and the disk are composed of ligulate florets, to the total exclusion of tubular ones. Such plants are called *Succory-headed* (Cichoraceæ), and are remarkable for their stems yielding a white milk, which, when concentrated, has a soporific quality. The other sections are destitute of this property. Here are arranged numbers of our wild flowers, such as *Goat's beard* (Tragopogon), *Sow Thistle* (Sonchus oleraceus), *Wall Lettuce* (Prenanthes), *Hawkweed* (Hieracium), and the shabby *Succory* (Cichorium Intybus), with its ragged leaves, and pale blue florets. It is also to this section that the *Lettuce* and *Endive* of the gardens belong.

From the last you will easily distinguish what is called the *Thistle-headed* section (Cinarocephalæ). These plants have no ligulate florets; all the florets are tubular, generally very wide at the mouth, and so much spreading beyond the involucre as to give the flower-head almost an hemispherical form. The leaves of the involucre are also, in most of the species, hard and spiny. It is in this section that you will find all the *Thistles, Saw-worts* (Serratula), and *Blue Bottles* (Centaurea); it is also remarkable for containing the *Artichoke* (Cynara Scolymus), the bottom of which you have probably often eaten without thinking much about Botany. The next time you have one on the table, remember that the *scales*, which you suck are the involucre, the *bottom* is the receptacle, and the *choke*, which is thrown away, is a collection of florets, separated from each other by numerous stiff hairs, growing out of the receptacle.

The third section has heads composed of both sorts of florets; tubular ones in the disk, and ligulate ones in the ray; hence they are called *Radiate* (Corymbiferæ). It sometimes happens that these have no ray, and then the young student might naturally confound them with the Thistle-headed section : but this need not be done if you remark that the florets of the Thistle-headed section are very wide in the mouth, and spread over the sides of the involucre, the scales of which are usually hard and spiny, while the florets of the Radiate section are narrow in the mouth, and not longer than the scales of the involucre, which are usually soft and leafy. A very little practice will soon prevent your falling into any such error. By far the greater part of Compound flowers belong to the Radiate section ; *Sunflowers, Asters, Daisies* (Bellis), *Chrysanthemums, Marigolds, Wormwood* (Artemisia), *Cudweed* (Gnaphalium), *Coltsfoot* (Tussilago), *Groundsel* (Senecio) and *Chamomile* (Anthemis), with thousands of others, for this section is of prodigious extent, form a most striking feature in the vegetable kingdom.

When we were talking of Umbelliferous plants I cautioned you against committing the error of supposing that all plants with their flowers in umbels belonged to that natural order. In like manner I must now explain to you that although the arrangement of florets in heads is universal in Compound flowers, yet that there are many plants in the world whose florets are placed in the same manner, but which do not belong to this tribe. For instance, *Eryngo*, which is an Umbelliferous plant, has its flowers in heads, so has *Sanicle*, another genus of the same order; and *Teasel* (Dipsacus) and *Scabious* (Scabiosa) are so extremely like Compound flowers in appearance, that you would never suspect them of being strangers of another family, if you were not apprized that no plants belong to the tribe of Compound flowers, which have not their *anthers united into a cylinder*. This it is which, taken with the disposition of the florets in heads, alone gives a positive character to the plants I have been speaking of. This remembered, you are mistress of the key to six or seven thousand species.

A few words upon the plant called Ribgrass, to which I have already alluded, and I pass at once to my promised analysis of the Monopetalous Dicotyledonous orders.

Ribgrass (Plantago) is a weed common at the foot of walls, by the side of pathways, and in moist places generally. It derives its name from its leaves having remarkably strong ribs passing from the bottom to the top: it has no apparent stem, but the leaves lie flat on the ground. The flowers (Plate XVII. 2. *fig.* 1.)

are white and green, disposed very closely in a long spike, and remarkable for their long stamens, the filaments of which soon become too weak to support the heavy anthers. Each flower has a hollow bract on its outside (*fig. 2. b.*), and a calyx of four green sepals, which are also concave, and overlie each other very much at the edges. The corolla is a thin, almost transparent, greenish-white body, divided at the end into four lobes (*fig. 2. c.*) which fall back on the sepals. Four stamens arise from the tube of the corolla, bearing on their filaments inverted arrow-shaped anthers. The ovary (*fig. 6.*) contains two cells, in each of which are many seeds (*figs. 7, 8.*), but it terminates in a simple hairy stigma; on this account it is considered to be formed of a simple carpel, notwithstanding its two cells. In course of time, the receptacle of the ovules separates from the sides of the ovary, and becomes covered all over with seeds of a pale chesnut colour (*fig. 10.*); at the same time the style drops off the ovary, which changes to a little hard dry brown case or capsule (*fig. 9.*), surrounded with the sepals, and separating transversely into two parts, by giving way at its base.

These characters entitle the Ribgrass to be considered the representative of a very distinct, although very small natural order, called the *Ribgrass tribe*; which is, however, of too little importance to make it worth detaining you longer about it.

Now let us again consider what the distinctions are of the Monopetalous orders, I have recommended to

your study.   Although they form only a part of what
really exist, yet a clear knowledge of them is a great
step towards that of the remainder.   I think the distinctions will be sufficiently well expressed in a tabular form, without any preliminary explanation,
beyond this that the number of the carpels, out of
which the ovary is formed, is the most important
fundamental distinction to employ.

* Ovary formed of more carpels than two.

Ovary split into four lobes	{ Flowers regular —*The Borage tribe.*† { Flowers irregular—*The Mint tribe.*†
Ovary not split into lobes	{ Erect bushes—*The Heath tribe.* { Climbing plants—*The Bindweed tribe.*

** Ovary formed of two carpels.

Ovary inferior { Milky plants { Anthers all separate—*The Harebell tribe.*
{ Anthers all united—*The Lobelia tribe.*
Not milky plants { Leaves without stipules—*The Honeysuckle tribe.*
{ Leaves with stipules—*The Coffee tribe.*

Ovary superior { Flowers irregular . . . *The Foxglove tribe.*
Flowers regular { Stamens two . . *The Olive tribe.*
{ Stamens five { Leaves opposite ribbed { *The Gentian tribe.*
{ Leaves alternate { *The Night-shade tribe.*

*** Ovary formed of only one carpel.

Anthers united in a cylinder . .	*Compound Flowers.*
Anthers distinct	*The Ribgrass tribe.*

---

† I hope the learned reader will pardon my having placed these two
orders among those whose ovary is formed of more carpels than two.
My reason for doing so is that they seem as if so constructed, and it
would be difficult to make a beginner understand that they are not.

EXPLANATION OF PLATE XVII.

I. COMPOUND FLOWERS.—1. Half a flower-head of *French Mari-gold* (Tagetes patula); *a* florets of the ray; *b* florets of the disk; *c* section of involucre; *d* receptacle; *e* flower-stalk—2. A floret of the ray; *a* ovary; *b* pappus; *c* tube, and *d* blade of the corolla.—3. Floret of the disk; *a* ovary; *b* pappus; *c* corolla.—4. Cylinder of stamens.—5. The same slit open and unrolled.—6. Half an involucre; *a* the receptacle.—7. The two stigmas.—8. A grain, ripe and crowned by the pappus *b*.—9. A section of a seed, shewing the embryo.

II. THE RIBGRASS TRIBE.—1. A spike of flowers of *Broad-leaved Ribgrass* (Plantago major).—2. A separate flower; *a* the calyx; *b* a bract; *c* the corolla.—3. A section of the calyx to shew the relative position of the sepals to each other; *b* the bract.—4. A sepal.—5. An anther with a part of a filament.—6. A pistil.—7. An ovary cut perpendicularly.—8. The same divided horizontally.—9. A ripe fruit invested by its calyx.—10. A cluster of seeds upon the receptacle, as they are left when the shell of the fruit falls off.—11. A seed.—12. The same cut through to shew the embryo.

LETTER XVIII.

DISTINCTIONS OF EXOGENOUS OR DICOTYLEDONOUS, AND OF ENDOGENOUS OR MONOCOTYLEDONOUS PLANTS—THE NARCISSUS TRIBE—THE CORNFLAG TRIBE.

(Plate XVIII.)

In my last lettter I took leave of Dicotyledonous plants, which I am persuaded you are by this time able to recognize by their general aspect, as well as by the technical distinctions to which I have chiefly called your attention. It is not their netted leaves, nor the concentric circles in their stems when woody, nor the two-seeded lobes of the embryo alone, by which they are known, but also other characters in combination with those. Their leaves are usually jointed with the stem, so that they are thrown off at certain seasons : in deciduous trees in the autumn, in evergreens in the spring or summer; their flowers, if perfect, or nearly so, are mostly divided by four or five; that is, have four or five sepals, and four or five petals, either distinct or combined ; and finally, their mode of growth is, in general, by branching repeatedly to form round-headed trees, or broad spreading bushy herbs. All these circumstances are characteristic of Dicotyledonous plants ; but most of them are occasionally subject to exceptions. When exceptions occur in regard to any one circumstance,

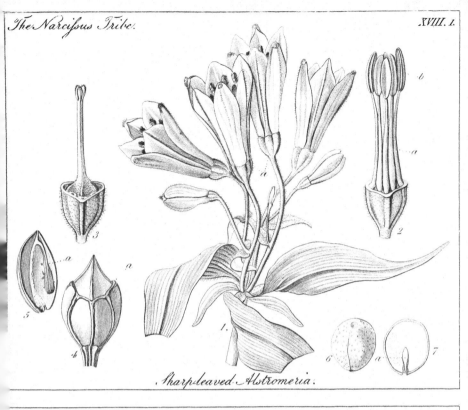

Sharp-leaved Alstromeria.

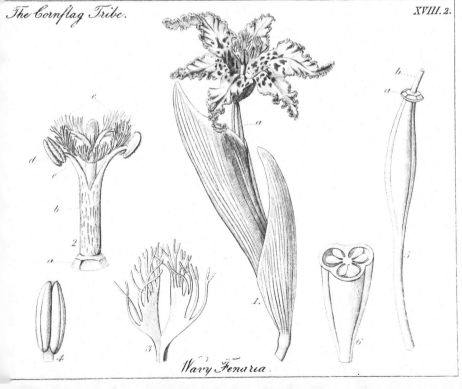

Wavy Fenaria.

attention should then be paid to others, for the purpose of ascertaining, whether, although a plant
may appear in some one point of structure not to
be Dicotyledonous, it may not upon the whole
possess the characters of that great division. For
instance, the common garden Pink, has leaves which
are not netted ; you might suppose on that account
that it is not Dicotyledonous; but when you find
that it is a branching plant, with five divisions of the
calyx, five petals, and ten stamens, you may be sure
that it is Dicotyledonous, notwithstanding the apparent deviation of the leaves from the general rule
of structure.

All this will be more clear to you, when you become acquainted practically with MONOCOTYLEDONOUS
plants ; which are quite different things. They often
shoot up into the air without any branches, and consequently have a sharp-headed appearance when
they form trees ; there is only one seed-lobe to their
embryo ; and their stem has no trace, whatever, of
concentric circles ; on the contrary, it presents, when
cut across, one uniform dotted surface, as you may
see if you take a piece of the common *cane*, which is
split for the bottoms of chairs, and which is in reality
the stem of a sort of Palm. Instead of enlarging
their stems by the addition of new wood to the outside of the old, Monocotyledonous plants only add
new matter to the centre of that which previously
existed, on which account they are named ENDO-
GENOUS (or growers inwardly). The veins of their
leaves run in nearly parallel lines from the base to

the point of the blade, without branching, or forming a kind of net work, as in the leaves of grasses or lilies ; so that in this respect they are immediately known from Dicotyledonous plants. Their flowers, moreover, are almost always divided by three, instead of by four or five ; you will, for instance, generally find three sepals, and three petals, and three, or twice three stamens, and an ovary made up of three car- pels ; so that there are abundant means of dis- tinguishing Monocotyledonous plants, whether you have only their stem, or their leaves, or their flowers, or even their fruit to examine. Now just observe, how important this is ; suppose you saw a simple-veined leaf of some plant, and nothing further ; although you might not be able to tell the name of the plant which bore the leaf, nor even its natural order, yet you would know that its stem must have grown by addition to its inside, that its embryo had only one seed-lobe, and that its flowers would in all probability be divided into three sepals, and three petals. Nor is this all ; nearly the whole of Monoco- tyledonous plants are harmless ; the chances would therefore be, that your leaf belonged to some harm- less plant ; and if it were the leaf of a tree, you would be perfectly certain that it came from some hot climate ; for no Monocotyledonous trees are found in cold countries. The inspection of a mere leaf would therefore lead you to a number of useful and interesting conclusions, at which you could never arrive if you studied Botany according to an artificial system.

As much of the facility of distinguishing Mono-cotyledonous from Dicotyledonous plants depends upon a familiarity with the appearance of their leaves, let me recommend you, before we proceed any further to procure those of Corn or Grass, of the Pine Apple, the Hyacinth, the Tulip, or the Daffodil, and compare them with any of the Dicotyledonous plants, you are now familiar with. In the mean-while I will sketch a comparative view of Monocoty-ledonous and Dicotyledonous plants, which you can afterwards study at your leisure.

DICOTYLEDONOUS PLANTS.	MONOCOTYLEDONOUS PLANTS.
Stems generally branched, and when old covered with cracked bark.	*Stems* generally quite simple; and when old covered with smooth bark.
Wood, consisting of concentric circles; the central part the oldest and hardest; the bark connected with a central pith, by means of thin plates called medullary rays.	*Wood*, not consisting of concentric circles; the central part the youngest and softest; the bark not connected with a central pith by means of medullary rays.
Leaves, with veins arranged in a netted manner; usually jointed with the stem.	*Leaves*, with simple parallel veins, which are not netted; usually not jointed with the stem.
Flowers, usually with the parts arranged in fours or fives.	*Flowers*, usually with the parts arranged in threes.
Embryo, with two or more seed-lobes, or cotyledons.	*Embryo*, with only one seed-lobe, or cotyledon.
Found wild as trees all over the world, except in the very highest latitudes.	*Found wild* as trees only in hot climates, and chiefly within the tropics.

These preliminary observations will, I trust, have conveyed to you some general notions of the nature

of Monocotyledonous plants ; among which you will
find that many of the commonest and most useful
tribes are arranged. You will scarcely, however,
have suspected that Lilies, and Palms, and Bananas,
with Hyacinths, Squills, Daffodils, and Orchises, were
associated with Grasses, Rushes, and Sedges ; the
natural affinity of all these is what we have now to
study. In order to entice my young friends onwards,
and to prevent their getting weary of their pursuit at
a time when I am most anxious that their attention
should be fixed, I shall begin with a natural order,
the species of which are so generally admired, that I
am sure they will wish to understand it botanically.

The *Daffodil*, called by Botanists Narcissus, represents a group of plants, having fine gay flowers,
and long narrow strap-shaped leaves arising from
bulbs which grow under ground. It is called in English the *Narcissus tribe*. We will not, however, take
tbe Daffodil to illustrate it, for reasons that I shall
mention by and bye. A beautiful Chilian plant,
called the *sharp-leaved Alstromeria*, of which I send
you a drawing (Plate XVIII. 1.), and which is now
not very uncommon in gardens, will shew you the
general structure of this tribe in a more satisfactory
manner.

Its leaves are of a firm and rather fleshy texture,
gradually taper to the point, and are filled with simple veins. It has an inferior ovary (*fig*. 1. *a.*), with
three angles, three cells, and many seeds in each
cell. From the upper end of the ovary rises a richly
coloured reddish-orange calyx of three sepals, inside

which are three orange-coloured petals, shaped exactly like the sepals; these are all rolled together so as to form a kind of bell. The stamens (*fig.* 2. *a.*) are six, and are remarkable for bearing blueish-purple anthers; a very uncommon colour for those parts. From the top of the ovary springs a three-cornered style, terminated by a three-lobed stigma (*fig.* 3.). Finally, a dull greenish-brown capsule is ripened, which is marked externally by six ribs (*fig.* 4.), and a horizontal line, which indicates the place whence the calyx, corolla, and stamens fell; when perfectly ripe it separates into three concave valves (*fig.* 5.), each of which carries away with itself a portion of the receptacle (*a.*) of the seeds. The seeds are spherical (*fig.* 6.), and consist of a great mass of albumen, in which lies a little cylindrical embryo (*fig.* 7.). This form of embryo is the most common in Monocotyledonous plants. Its upper end is the cotyledon, its lower the radicle, and the plumule or rudiment of a stem lies concealed in the former.

Of these characters a part is confined to the genus Alstromeria, and a part peculiar to the Narcissus tribe. The latter is briefly this : *an inferior ovary, calyx and corolla of the same form, and six stamens (or more).*

To this definition answers the Daffodil in all respects ; but it has, in addition, a long cup, which is a row of abortive stamens ; and this circumstance marks its genus. Some very fragrant hot-house plants, called Pancratiums, or *Sea-side Lilies,* also have a cup, but their anthers arise from its border, which at once distinguishes them from the Daffodil.

It is to this order that also belong the golden *Sternbergia* of autumn, the scarlet *Amaryllises* of Brazil, the *Belladonna Lilies* of the Cape, and *Crinums*, with their long and relaxed petals, sometimes of the French white, and sometimes deeply stained with crimson. *Snow-drops*, too (Galanthus), and *Snow-flakes* (Leucojum), whose names so well express their colour, the ruby-petaled *Nerines*, to which belongs the *Guernsey Lily*; and *Blood-flowers* (Hæmanthus), whose juice is a mortal poison, are all allies of the Narcissus.

We have often seen that venomous properties lurk beneath the fairest forms, and that external appearance offers no beacon to warn the traveller of the plants in which danger lies concealed. The Narcissus tribe affords another instance; the bulbs of the Daffodil are emetic, those of the Blood-flower yield a deadly gluey poison, with which the African savages smear their arrow-heads, and the bulbs of the whole tribe are suspicious. Fortunately, however, its botanical characters are so precise, that there is no difficulty in distinguishing it from all others.

So like is a *Crocus* to some of the Narcissus tribe, that a student would naturally suppose it to belong to it, especially when he found that it also had an inferior ovary, with three cells, and its sepals and petals so much alike, as to be distinguishable only by one being rather differently coloured, and placed on the outside of the other. It differs, however, in having only three stamens instead of six, and in the anthers being turned with their faces towards the sepals, instead of towards the style ; a singular peculiarity, which in this case is found to indicate a total absence of poisonous properties ; so that while the Narcissus tribe is dangerous, the natural order, called *the Cornflag tribe*, to which the Crocus belongs, is perfectly harmless.

These plants vary a great deal in their general appearance, owing to the different shapes of the calyx and the corolla, and to the size of the stem, which is sometimes round and subterranean, as in the Crocus, and sometimes long, scarred, and creeping on the surface of the ground, as in the *Cornflag* or Iris ; of which the *Orris root* which babies suck is a preparation, and the name a corruption. They may, however, be always known by their agreeing with the Narcissus tribe in every thing except their *three stamens with the anthers turned away from the style*, and the want of bulbs. Their leaves are also very unusually thin, and shaped like a straight sword-blade, with the edge turned to the stem, on which account they used to be called *Ensatæ*, or sword-leaved plants (Plate XVIII. 2. *fig.* 1.).

Along with the drawing of the Alstromeria you have another, which represents a most singular plant, the colour of whose flowers resembles a lizard's back. It is a native of the Cape of Good Hope, and called the *Wavy Ferraria* (Ferraria undulata). Its flowers consist of three sepals and three petals, all of which are so wavy and curly at their edges, and so much

alike that you can hardly distinguish them ; below a
kind of cup, formed by the union of these parts, is a
long ribbed ovary (*fig.* 1. *a.*), which contains three
cells and many seeds (*fig.* 6.). The three stamens are
in this genus grown into a column (*fig.* 2.) like the
column of a Passion-flower ; but in other genera they
are distinct ;   each at its point curves away from the
stigma as if to convey its anther (*fig.* 2. *d.*) as much
as possible out of the reach of it, averting from it
its face (*fig.* 2. *d.*).   The style is a small cylinder,
divided at its end into three broad lobes (*fig.* 2. *e.*)
each of which is separated into two parts (*fig.* 3.)
cut up into five hair-like segments : these lobes are
the stigmas.

In Iris, the genus from which the tribe takes its Latin
name (*Irideæ*), the structure is more curious than in
Ferraria ; the three sepals are broad and spreading,
and often ornamented with a beautiful feathered crest;
the three petals stand erect, and curve over the
centre of the flower ; while the stigmas are broad
richly coloured parts, resembling petals, and curve
away from the centre, as in the Ferraria.   At first
sight you would suppose the Iris was altogether destitute of stamens ; but if you lift up the stigmas you
will find the runaways snugly hidden beneath their
broad lobes, and lying close to a humid lip through
which the influence of the pollen is conveyed to the
ovules.   This widening of the stigma is a very common event in the Cornflag tribe ; even in the Crocus
it occurs, only the stigmas are so rolled up that you
do not discover it until you unroll them ; they are,

in fact, so much heavier than the power of the style can support, that in the Saffron Crocus, in which they constitute the substance called Saffron, they hang down on the outside of the flower like an orange tassel.

It is to this natural order that belong those countless species of Ixia, Gladiolus, Watsonia, Babiana, &c. which spring up at the Cape of Good Hope upon the commencement of the rains, and soon cover the parched and half naked karroos with a robe of the deepest green, adorned with all manner of gay and sparkling colours.

Having studied these plants in the order I have mentioned, compare them in your mind with the numerous Dicotyledonous tribes you have already become acquainted with, and see whether you can anticipate any difficulty in recognizing other tribes of Monocotyledonous plants, by the external characters I have pointed out; without the slightest necessity for having recourse to the examination of their stems or seeds.

EXPLANATION OF PLATE XVIII.

I. THE NARCISSUS TRIBE.—1. A shoot of the *sharp-leaved Alstromeria* (A. acutifolia); *a* the inferior ovary.—2. An ovary deprived of calyx and corolla, but with the stamens in their place.—3. An ovary without even the stamens; shewing the style and stigmas.—4. A ripe fruit of Alstromeria psittacina *(the Parrot Alstromeria)*, just before it separates into valves; *a* the scar whence the stamens, calyx, and corolla dropped.—5. One of the valves, with a part of the receptacle of the seeds *a* adhering to it.—6. A seed.—7. A section of the same ; exhibiting the embryo *a* lying in the midst of the albumen.

II. The Cornflag Tribe.—1. A piece of *the Wavy Ferraria* (F. undulata); *a* the ovary.—2. A column of stamens; *a* the base of the sepals and petals; *b* the column; *c* the fore part of the filaments ; *d* the anthers; *e* the stigmas.—3. One of the stigmas separated from the style.—4. An anther.—5. An ovary; *a* the place whence the sepals and petals have been removed; *b* the base of the column of stamens.—6. A horizontal section of the ovary.

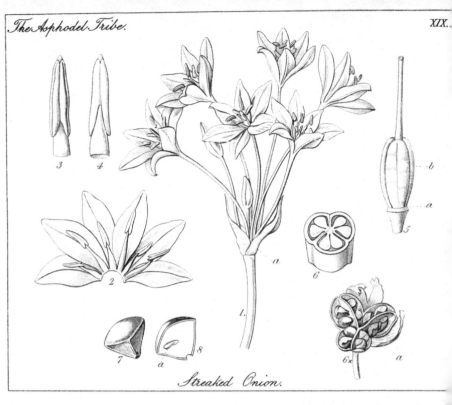

Streaked Onion.

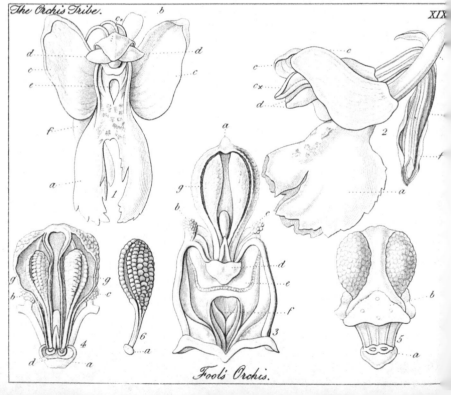

Fool's Orchis.

LETTER XIX.

THE ORCHIS TRIBE—THE ASPHODEL TRIBE.

(Plate XIX.)

THE last natural order of Monocotyledonous plants, with an inferior ovary, which I propose to mention to you, is the *Orchis tribe,* one of the most extensive and curious of the Vegetable Kingdom. In the end of May, or beginning of June, you will meet with great numbers of a fine common species in the pastures and meadows, where they rise with their spotted leaves, and spikes of purple or rosy speckled flowers, conspicuous among the herbage that surrounds them. The country people call them Cuckoo-flowers, because they make their appearance when the cuckoo begins to call. Of these the commonest of all are the *Male Orchis* (Orchis mascula), and *Fool's Orchis* (O. Morio), either of which you may gather for the purpose of examination.

Suppose we take Fool's Orchis. It has a little fleshy root, composed of two oval knobs, and a number of succulent fibres ; its leaves spread at the surface of the ground, are of a light green, of a narrow strap-shape, tapering to a blunt point, with a soft, rather fleshy texture, and the usual simple-veined structure of Monocotyledonous plants. The stem is erect, and about nine inches high, sometimes less, but seldom more; it is covered by two or three leaves wrapped round it, so as to form a sort of sheath; at its top it bears a spike of flowers, coloured with green and white, and having the following most remarkable structure.

From the bosom of a narrow wavy lance-shaped bract (Plate XIX. 2. *fig.* 2. *f.*), springs an angular twisted body, pure green at its lower end, but often stained with red near the top (*fig.* 2. *e.*); this is an ovary, containing only one cell, and an inconceivable multitude of minute ovules, arranged in double rows upon three narrow receptacles; at its point it is gracefully bent forward, as if to present you with its singular green and pink flower.

The first thing that usually strikes an observer, is a broad roundish notched leaf (*fig.* 1. and 2. *a.*), hanging from out of a sort of casque or helmet, the two sides of which expand and retreat a little (*fig.* 1. *c. c.*), so that they may set off the lively rose colour and rich crimson blotches with their own dull green stripes; this leaf is called the LIP (labellum), and separates a little into three divisions, of which the side ones are the most notched, and the largest; at its base is a hole (*fig.* 1. *e.*), which will prove, upon examination, to be the mouth of a rose-coloured horn, that seems to swing lightly behind the flower (*figs.* 1. and 2. *b.*); it is usually called the SPUR. The other leaves of the flower are thus: two are concave, spreading, a little turned in at the point, and charmingly striped with green veins, both within and without (*figs.* 1. & 2. *c.*); another stands quite at the back, coloured with pure rose, and projecting forward over the lip (*figs.* 1. & 2. *c**.); below this lie two other leaves, narrower, and more delicate than the last, forming, together with it, the casque out of which the lip seems to hang (*figs.* 1. & 2. *d. d.*). These are the parts which answer to the calyx and corolla in other plants; at first sight, you will be at a loss how to determine which belong to the one, and which to the other; and especially to know what the lip is with its spur; look, however, more attentively, and remark, firstly, that the whole number of leaves in the flower is *six*; and secondly, that *three* of them, namely, the lip, and two narrower and more delicate leaves, which form part of the casque (*figs.* 1. & 2. *a.* & *d. d.*), are placed within the three others. The first of these (*c. c.* and *c**.) form the calyx, and the last (*a.* and *d. d.*) the corolla; a very irregular one, certainly, but nevertheless conformable to all the rules of organization; as for their irregularity, *that is characteristic* of the Orchis tribe. A little study and examination by yourself, will satisfy you that this is the true view of their nature, and all the strange appearance which puzzled you at first, was caused by the unusual manner in which the sepals and petals are shaped and directed, and the disproportionate size of one of them.

The next object that is to engage your attention, is far more singular and difficult to understand. In the centre of the flower, in the place of stamens and style, just at the back of the hole that leads down

into the spur, half hidden by the petals, stands a
crimson flat body (*fig.* 3.) having a deep furrow in
its front.   I should quite despair of explaining the
structure of it, without the assistance of a drawing,
which I accordingly send you : the parts are a good
deal magnified, as is usual in the dissections of flowers by Botanists, in order to bring every thing distinctly into view.   If you look attentively at the front
of the central body (*fig.* 3. *a.*), you will remark, in
the first place, that it is separated into two lobes, by a
deep channel drawn down its middle ; and secondly,
that each lobe will open, if pressed, by a suture (*fig.*
3. *g.*) running through it from one end to the other.
Pull asunder the two sides of each lobe, so as to lay open
their inside (*fig.* 4.), and in each there will be seen
an olive-green granular mass (*fig.* 4. *g.*), tapering
gradually into a thin stalk, at the end of which is a
viscid semi-transparent gland (*fig.* 6. *a.*).   If you
squeeze a portion of the olive-green granular mass in
water, beneath a very powerful microscope, you will
be surprised to see that it consists of infinite multitudes of grains of pollen sticking together in threes or
fours ; it is therefore a mass of pollen in a singular state ; and this fact being taken as the basis of
your reasoning as to the nature of the other parts
connected with it, it will result as a necessary consequence, that the central body is an anther, and its
sutures the lines of dehiscence, or of opening.   At the
foot of the anther is a pale whitish-green fleshy cup
(*fig.* 3. *d.* and 4. *a.*), in which the glands at the end
of the stalk of the pollen masses are concealed ; this,

which is botanically called the HOOD, or pouch (cu-cullus, or bursicula), is peculiar to some of the Orchis tribe, and is caused by a doubling upwards of the upper edge of the stigma.

The stigma itself is a broad viscid shining space (*fig.* 3. *e.*) lying just below the hood, between it and the mouth of the spur.

These things being thus made out, it follows, you see, that the column of an Orchis is a body formed of a stamen, a style, and a stigma, all grown into one solid body; and this is the great peculiarity of the Orchis tribe. Its genera vary amazingly in the structure of the anther, the column, the lip, and in-deed of all the parts, but in *the consolidation of the style and stamen*, they are all agreed. This then is the characteristic of the Orchis tribe.

If, however, there was really only one stamen, this curious natural order would be more at variance with the usual structure of Monocotyledonous plants than its conformity in the calyx and corolla would lead one to expect; and accordingly we find, that although only one stamen is perfect, there are dis-tinct traces of two others in an extremely imperfect state. On each side of the anther, near its base, you will find a roundish granular knob (*fig.* 3. *b. c.*), which has been ascertained to be the rudiment of another anther; so that in reality the column is composed of one perfect stamen, standing between two imperfect ones; a most striking proof of the harmony of design which is manifested in all these Monocotyledonous tribes.

If you consider the mutual relation which all the parts
of the Orchis bear to each other, you will scarcely fail
to be struck with one circumstance above all others,
namely, the apparent want of any means of communication between the pollen and the stigma. Not
being in fine powder the pollen is not able to be scattered in the air, like that of other plants; if it were to
fall out in a mass it would hardly touch the stigma;
and even the possibility of this seems to be purposely
prevented by the glutinous gland to which the stalk
of the pollen mass adheres, and which is itself confined within the pouch. To account for the manner
in which the necessary contact between the stigma
and the pollen takes place, two explanations have
been given; one that insects, inserting their pro-
boscis into the flower in search of honey, disturb
and pull out, or unintentionally carry away with
them the pollen mass which sticks to them by means
of the gland, and that in the latter case, buzzing from
flower to flower they leave it behind them on the
stigma of some neighbouring blossom; the other explanation, which has been offered by the celebrated
Mr. Bauer, is that the influence of the pollen is not
communicated to the stigma, by actual contact of the
pollen, as is usual in other cases, but that it passes
down the stalk of the pollen mass, into the gland,
and thence to the humid surface of the stigma; and
he has shewn that great probability attaches to this
opinion, in consequence of the existence in such plants
as the Orchis itself, of a beautiful contrivance to
secure such a communication. He discovered that in

the bottom of the pouch, just below the glands, are two little passages (*fig.* 4. *d.*), which open directly over exactly that part of the stigma which appears the best adapted for receiving the influence of the pollen : and that hence the pouch, which seems to the superficial observer a means of preventing communication between the pollen and the stigma, is in fact a most admirable contrivance of nature to ensure it. I leave you to form your own opinion, as to which of these two explanations is the more plausible.

In this country the more common or remarkable of the wild plants belonging to the Orchis tribe, are the *Bee Orchis* and *Fly Orchis* (Ophrys), whose lips resemble the insects after which they are named, the *Man Orchis*, and the *Lizard Orchis* (Aceras), with their yellow or purple strap-like lips, *Neottia*, with her russet flowers, and scaly stems, springing from a cluster of entangled roots, which have given her the name of *Bird's Nest*, the *Butterfly Orchis* (Platanthera) with its long and taper cream-coloured lip, and the *Ladies Slipper* (Cypripedium), with its large yellow bladdery flowers, and twin anthers. They all grow on the ground in meadows, or marshes, or woods ; and have always been considered the most curious and beautiful plants of a European Flora. But it is in tropical countries, in damp woods, or on the sides of hills in a serene and equal climate, that these glorious flowers are seen in all their beauty. Seated on the branches of living trees, or resting among the decayed bark of fallen trunks, or running over mossy rocks, or hanging above the head of the admiring traveller, suspended from the gigantic arm of some monarch of the forest, they develope flowers of the gayest colours, and the most varied forms, and they often fill the woods at night with their mild and delicate fragrance. For a long time such plants were thought incapable of being made to submit to the care of the gardener, and Europeans remained almost ignorant of the most curious tribe in the whole vegetable kingdom. But it has been discovered of late years that by care and perseverance they may be brought to as much perfection in a hot-house as they acquire in their native woods, and they now, under the name Orchideous Epiphytes, form the pride of the collections of England.

Unfortunately they are of scarcely any known use. *Vanilla*, which you eat in creams and other sweets, is the pod of a kind which, in the West Indies, creeps over trees and walls like ivy; and a nutritive substance called Salep is prepared from the tubers of a kind of *Eulophia*; a most meagre catalogue of useful properties in a tribe of near two thousand species, but one to which nothing can be added, if we except what is called the *Shoemaker plant* (Cyrtopodium Andersonii), whose stems afford a glutinous extract employed by the Brazilians for sticking together thin skins of leather.

This order, like the two preceding has, as we have seen, an inferior ovary, a character by which they, and several others which I have not time to mention, are readily known. In your study of Monocotyledonous plants you cannot do better than take this circumstance as a fundamental distinction from which your analysis may be continued. Let us now proceed to some orders in which the ovary is superior.

The *Hyacinth, Squill, Onion, Star of Bethlehem* (Ornithogalum) and *Asphodel*, belong to a natural order called the *Asphodel tribe*, which is remarkable for the extreme simplicity of the structure of all its parts. Three sepals, and three petals of similar form, size, and colour, six stamens, and a superior three-celled ovary, which changes to a fruit containing seeds with a black brittle skin (Plate XIX. 1.), form the essential character, and combine a large number of plants, generally quite harmless, and in the majority of cases remarkable for either their use or their beauty. It is difficult to single out any one species better fitted than another to illustrate the Asphodel tribe, so uniform are they in the more important points of organization. I happen to have at hand the *Streaked Onion* (Allium striatum), but you may take with equal advantage the common Onion, or any of the others I have above named, for the purpose of study.

The leaves and flowering stems of this plant, rise from a subterranean roundish fleshy body, formed of scales wrapped closely over each other. The scales are of the same nature as those of a bud, namely, the rudiments, or the bases of leaves; and the body itself, called a BULB, is a kind of underground bud; hence you will perceive that when one talks of Hyacinth *roots* which are placed in glasses, or of the *roots* of Onion, Garlic and Shallots, an incorrect kind of language is made use of. In one respect, the bulb differs essentially from a bud : it is not a perishable part, intended merely as a protection to the young and tender vital point, from which new growth is to take place : this indeed is a part of its object, but it also serves as a copious reservoir of nutritive matter upon which the young leaves and flowers feed. On this account its scales are not thin and easily withered up, as in a common bud, but succulent, and capable of retaining their moisture during long and severe drought. In this we again see a direct manifestation of the all protecting care of the Deity ; for bulbous plants are generally natives of situations which at certain seasons of the year are quite dried up, and where all vegetation would perish if it were not for some such provision as we find in the bulb ; in places like the hard dry Karroos of the Cape of Good Hope, where rain falls only for three months in the year, in the parched plains of Barbary, where the ground is rarely refreshed by showers, except in the winter, and on the most burning shores of tropical India, beyond the reach of the tide, and buried in sand, the temperature of which often rises to 180°, bulbous rooted plants are enabled to live, and enliven such scenes with their periodical beauty.

You must not, however, imagine, either that all the Asphodel tribe have bulbs, or that all bulbs belong to the Asphodel tribe ; of the inaccuracy of the latter notion you must already be aware, if you remember the Narcissus tribe ; the former would be not less a mistake ; all that I meant, in thus connecting the bulbous structure with the Asphodel tribe, was that it is an exceedingly common characteristic. The Asphodel genus itself has a fleshy fingered root,. without any trace of bulb, and some of the genera contain trees of considerable size.

The leaves are long narrow things, like green straps, and have the simple parallel veins you have been led to expect. The flowering stalk rises directly from the bulb, without any intermediate stem ; as it is long, and destitute of leaves, and rather different in appearance from a common flower-stalk, it is technically named the SCAPE. At its top it bears an umbel of flowers, at the base of whose long stalks are a number of membranous satiny scales, or bracts (*fig.* 1. *a.*) ; they are a sort of involucre, but are occasionally called a SPATHE. The three sepals are very evidently placed on the outside of the petals (*fig.* 2.), but excepting in this respect, they are absolutely the same both in colour, size, form, and direction. Of the six stamens, three are a little smaller than the others ; their anthers open by two slits which are turned towards the style. The ovary (*fig.* 5.) is an oblong body, with three furrows, a single style, and a stigma, which exhibits no sign of being divided. Inside the ovary you will find three cells, in each of which is a number of ovules (*fig.* 6.). Last of all comes the fruit ; a little brown dry case which splits into three valves (*fig.* 6*.), to allow of the escape of the angular black seeds, whose skin is of a very brittle nature. This last circumstance of the black

brittle coating of the seed, is one of the most important characters in the Asphodel tribe, as you will find
in my next letter.

The principal differences of appearance in the
Asphodel tribe are caused by two circumstances, the
growing together of the sepals and petals into a tube,
and the formation of a stem covered with leaves; the
former alters the look of the flowers, the latter
changes the whole aspect of the plant. You must
therefore pay attention to this, before you discontinue
your study of the tribe.

To understand the first of these two circumstances
you should endeavour to trace the gradations by
which the growing together of the sepals and petals
occurs. In the Onion you have seen that they are
all distinct; they are equally so in the *Vernal Squill*
(Scilla bifolia), or in the *Nodding Star of Bethlehem*
(Ornithogalum nutans); but in the wild *Blue Hyacinth*, or *Blue-bell*, as it is often called (Hyacinthus
non scriptus), they converge so as to form a tube;
and in the curious *Starch Hyacinth* (Muscari) they
are completely glued together. These gradations
exemplify most perfectly the passage from a spreading flower with separate segments to a tubular flower
with all the segments united.

The second peculiarity, that of forming a leafy
stem, gives a much more different aspect to a part of
the tribe. In the *Asparagus*, for example, which you
perhaps only know in the state in which it is brought
to table, the stem when full grown is repeatedly

branched and covered with little taper green leaves ; the *Asphodel* itself has a simple stem clothed with very long blueish-green channeled leaves ; and certain exotic plants called *Dragon trees* (Dracænas), form trees of considerable size, with stems having tufts of long broad leaves at their ends, in the manner of Palms.

The latter, however, you are not likely to meet with, unless you should travel into countries more southern than Europe ; so that I do not anticipate any probability of your being embarrassed by them in your notions of the Asphodel tribe.

Our next subject will be Lilies, the most gorgeous, and Rushes the most mean, of Monocotyledonous plants.

EXPLANATION OF PLATE XIX.

I. THE ASPHODEL TRIBE.—1. An umbel of the *Streaked Onion* (Allium striatum) ; *a* the spathe.—2. A flower spread open.—3. An anther viewed in front.—4. The same viewed from behind.—5. A pistil ; *a* the place where the sepals and petals were cut off ; *b* the ovary.—6. An ovary cut through horizontally.—6*. A ripe capsule, separated into valves ; *a* the remains of the flower.—7. A seed.—8. The same cut through, shewing the embryo, *a*, lying in a cavity of the albumen.

II. THE ORCHIS TRIBE.—1. A flower of *Fool's Orchis* (O. Morio) seen in front ; *a* the lip ; *b* point of spur ; *c c* lateral sepals ; *c** upper sepal ; *d d* petals ; *e* mouth of the spur ; *f* bract covering the ovary.—2. The same viewed in profile ; *a* lip ; *b* spur ; *c c* lateral sepals ; *c* upper sepal ; *d* petals ; *e* ovary ; *f* bract enveloping the ovary.—3. Column of Orchis Mascula ; *a* the anther ; *b c* abortive anthers ; *d* pouch of the stigma ; *e* the glutinous face of the stigma ; *f* the mouth of the spur ; *g* the suture by which the lobe of the anther opens.—4. An anther with its lobes forced open ; *a* pouch; *b c* abortive anthers ; *d* passages in the pouch through which the pollen is supposed to communicate with the glutinous surface of the stigma ; *g g* pollen masses.— 5. Back of an anther ; *a* the passages in the pouch of the stigma ; *b* the part where the anther was cut off.—6. A pollen mass ; *a* the gland.— (These are after drawings by Mr. Bauer.)

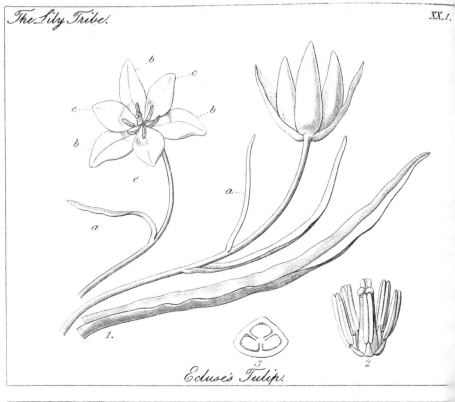

Ecluse's Tulip.

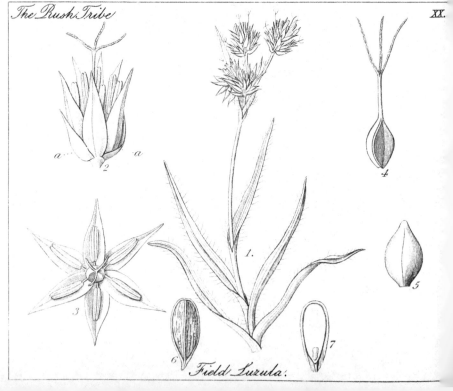

Field Luzula.

LETTER XX.

THE LILY TRIBE—THE COLCHICUM TRIBE—
THE RUSH TRIBE.

Plate XX.

THE characters of the Asphodel tribe are so simple, that I see no probability of your misunderstanding their application in more than two cases. Let your next inquiry then be how to avoid committing those errors.

If you examine the flower of a *Tulip* (Plate XX. 1.), you will find so great a resemblance between its structure and that of the Asphodels, that you will not doubt its belonging to their tribe. It has three sepals (*fig.* 1. *b. b. b.*), which are of the same size and colour as the three petals (*c. c. c.*); from within these arise six stamens (*fig.* 2.), and in the centre of all is an ovary with three angles, three cells (*fig.* 3.), and a number of ovules in each cell. All this is quite in accordance with the structure of the Asphodel tribe. The leaves, too, are extremely similar; they are narrow strap-shaped things, with simple parallel veins; the underground bodies, moreover, from which the leaves and flowers shoot up, are true bulbs. Surely, then, the Tulip must belong to the Asphodel tribe.

If the same accordance existed in the structure of all their other parts, it is clear, that to distinguish the Tulip from the Asphodel tribe would be unnecessary ; but there are distinctions in parts to which I have not yet adverted, and these distinctions are considered important.

In the first place, the flowers of the Tulip are *much larger, and more shewy* than you find in the Asphodels ; and secondly, its seeds have not the same hard black brittle coat ; but in its room you have *a soft pale spongy integument.* These are the two points upon which Botanists rely for the separation of the Tulip and its relations.

It is to the *Lily tribe* that the Tulip belongs, together with the speckled *Fritillaries, Crown Imperials, Day Lilies* (Hemerocallis), and true *Lilies* (Lilium), *Dog's-tooth Violets* (Erythronium), and *Tuberoses* (Polyanthes), the most odoriferous of flowers.

That the Lily tribe may be easily distinguished by the large size of its flowers, must be obvious, after naming such well known plants as the above, compared with which, the Asphodels, pretty as they are, sink into comparative insignificance. But the difference would be still more striking, if we added to the list the gigantic Lilies of Nipal, one of which is described by Dr. Wallich as growing ten feet high, with flowers large in proportion.

It is, however, certain, that notwithstanding the distinctions I have pointed out, the Lily tribe is very closely allied to that of Asphodels, with which it also coincides in its harmless qualities. We only know its beauty; it is in the eastern parts of Asia, among the Kamtchadales, &c. that it is applied to useful purposes. By those people, the bulbs of certain Lilies are used as a common food, and are stored up as an important part of their winter stock of provisions.

The fibres of the leaves of both Asphodels and Lilies are sometimes strong enough to be manufactured into hemp. Among the former is classed the *New Zealand Hemp* (Phormium), of which such extensive use is now made in the navy; and with the latter are stationed the plants called *Adam's Needle* (Yucca), because their strong sharp-pointed leaves have been fancifully compared to a gigantic needle. A better name would have been *the Needle-and-thread Plant*, for by soaking in water, the fibres of the leaves may be separated from the pulp, without being torn from the hard sharp point, so that, when properly prepared, the leaves do really become needles, ready provided with a skein of thread.

Since both the Asphodel and Lily tribes are so generally harmless, it becomes highly important that all possibility of confounding them with a poisonous order, which they in some respects resemble, should be guarded against. *Meadow Saffron*, called by Botanists Colchicum, *White Hellebore* (Veratrum), and some other plants, have a structure very analogous to theirs. A calyx and corolla, each of three leaves of similar form and texture, half a dozen stamens, and a superior three-celled ovary, also characterise in part the natural order called the *Colchicum tribe*, to which those plants belong. Colchicum itself is very like a Crocus in flower, but its superior ovary prevents its being confounded with the tribe in which the Crocus is included. The species of Melanthium and Helonias are so similar in appearance to many of the Asphodel tribe, that they would no doubt be referred to the latter by a young Botanist. They, however, Meadow Saffron, and all the rest of the Colchicum tribe, may be recognised at once by three marks; in the first place, they have no bulbs, but in their stead a solid knob or subterranean stem; secondly, their anthers are turned away from the stigma, splitting, and emitting their pollen on the side next the petals; and lastly, the three carpels out of which the three-celled ovary is constructed, are separated at their points, so that there are always three styles instead of one style. These signs are what you must trust to in your determination of the Colchicum tribe; they may appear slight, and you may wonder why such trifling distinctions should serve to distinguish poisonous from wholesome tribes; but with considerations of the causes of such a fact, we have no concern; all that it imports us to know is, that Providence has distinguished them by such minute marks, and has thus provided man with safe and unerring guides, if he will but learn how to follow them.

Between many other tribes of Monocotyledonous plants the distinctions are no stronger than those we have already examined, a proof of which I am now about to give you.

Rushes have so little apparent resemblance to Lilies, or Asphodels, or Meadow Saffron, that no one would ever dream of placing them all in the same natural group; and yet, if you examine a Rush, you will be surprised to see how slight is the difference that really exists between them. Some Rushes are humble rigid leafless herbs, with stiff slender wiry stems, and little clusters of dingy flowers; others are still more dwarfish in stature, but have well formed leaves. An exceedingly common plant of the latter description is the *Field Luzula* (Luzula campestris, Plate XX. 2.). From a fibrous root rises a stem not more than five or six inches high, and which you can hardly distinguish from the grass that it generally grows among in the meadows. Its leaves are narrow grassy things, clothed with remarkably long hairs. At the top of the stem grow a few heads of chesnut-brown flowers, the structure of which you will hardly make out, except the sun shines, for it is then chiefly that the parts unfold. But if at that time you watch one of the brown clusters, you will be able to perceive, that each flower has six chesnut-coloured leaves, which spread like a star (*fig.* 3.), of which three are sepals, and three are petals. On their outside are two or three bracts (*fig.* 2. *a. a.*), so like them, as to be only distinguishable by their position. From within the flower rise six stamens; and from between the latter an ovary (*fig.* 4.) with three angles, one style, and three stigmas. Its fruit (*fig.* 5.) is an ovate body, containing only one cell, and three seeds (*fig.* 6. 7.), with a pale soft skin.

R

Now you will observe that in all that relates to the sepals, petals and stamens, there appears to be nothing to distinguish a Rush from the Asphodel tribe ; the fruit would seem essentially different, because of its having only one cell ; but in the *true Rushes* (Juncus) the fruit has three cells, so that that difference is unimportant. In fact if we neglect the texture, colour, and imperfect degree of developement of the calyx and corolla, there will be hardly any means of separating Rushes from Asphodels, except the softness of the skin of their seeds. That the former do represent an inferior order of vegetation, there can, however, be no reasonable doubt ; we accordingly consider them among the most imperfect of regular Monocotyledons, Asphodels being an intermediate degree of developement, and Lilies the highest degree.

Rushes are of so little use in any of the ordinary affairs of life, that it is scarcely worth occupying more time with their study. The plants from whose stems are made what are called rush-bottomed chairs, rush-mats, and the like, are usually species of *Club-Rush*, and belong to the Sedge tribe, to which we shall now come very shortly,

The orders of Monocotyledonous plants which we have examined, and several others that have not been mentioned, form what may be considered a natural subdivision, characterised by the perfect and complete manner in which their flowers are organized. In no case do they exhibit fewer than six divisions, of which three belong to the calyx, and three to the corolla. If you examine with care the distinctions by which they are separated from each other, you will find that they are of this nature.

Ovary inferior
- Stamens and style separate
  - Stamens 6—*The Narcissus tribe.*
  - Stamens 3—*The Cornflag tribe.*
- Stamens and style grown together into a solid column . . *The Orchis tribe.*

Ovary superior
- Anthers with their faces turned towards the ovary. Carpels firmly united.
  - Seed-coat, soft and pale. Flowers large — *The Lily tribe.*
  - Seed-coat black and brittle. Flowers middle sized — *The Asphodel tribe.*
  - Seed-coat soft and pale. Flowers minute, brown and dry — *The Rush tribe.*
- Anthers with their faces turned away from the ovary. Carpels partially separate — *The Colchicum tribe.*

EXPLANATION OF PLATE XX.

I. THE LILY TRIBE.—1. Leaf and flowers of *Ecluse's Tulip* (Tulipa Clusiana); *a* bracts; *b* sepals; *c* petals.—2. Stamens and ovary.—3. The ovary cut through horizontally, shewing the ovules.

II. THE RUSH TRIBE.—1. Stem, leaves, and flowers of the *Field Luzula* (Luzula campestris).—2. A flower separate; *a a* bracts.—3. An expanded flower seen from above.—4. A pistil, with its single style and three stigmas.—5. A ripe fruit.—6. A seed.—7. The same cut through perpendicularly.

LETTER XXI.

THE BULLRUSH TRIBE—THE ARROW-GRASS TRIBE —THE DUCKWEED TRIBE.

(Plate XXI.)

In my last letter I mentioned that all the orders we have hitherto examined among Monocotyledonous plants have their flowers organized in a complete and perfect manner. This letter will relate to some that are incomplete and imperfect. You will find that there are three degrees of organization in Monocotyledons, of which we have passed over one, are about to enter upon a second, and shall very soon arrive at the third.

Bullrushes (Typha) are narrow flat-leaved tall plants, growing in marshes or pools of stagnant water, having their stems terminated by a dark cylinder, surmounted by a more slender yellow spike. These plants represent a natural order, named after them the *Bullrush tribe,* to which belongs a common wild marsh plant, called *Bur-reed* (Sparganium). The latter will furnish you with the means of studying the peculiarities of the natural order.

If you regard the Bullrush ever so attentively, you will fail to discover, at any period of its growth, a

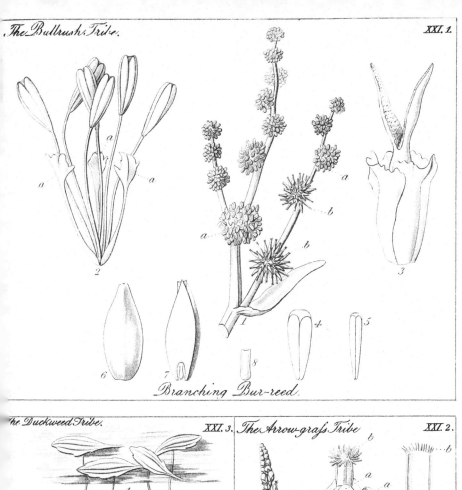

Branching Bur-reed.

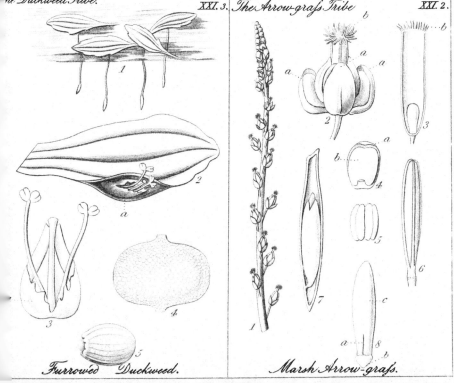

Furrowed Duckweed.   *Marsh Arrow-grass.*

trace of flowers like those of the plants that have gone before. Upon the whole surface of the plant you find nothing different from leaves, except the dark cylinder, and the yellow spike on the top of it. It is at these parts that in reality the flowers are collected; but they are so minute, and their organs are so delicate, that it requires a microscope, and a careful separation of them, to determine their real nature. The dark cylinder consists wholly of flowers containing pistils only, or being fertile, and the yellow spike is composed of flowers containing nothing but stamens, or being sterile. I do not, however, propose to fatigue you by obliging you to anatomize a Bullrush; the same purpose will be effected, if you take the *Bur-reed*, which is similar in structure, but larger in all its parts.

Branched Bur-reed (Sparganium ramosum, Plate XXI. 1.), is a wild plant, frequently found growing in ditches or pools, or by the wet banks of rivers. It rises to the height of two, or even three feet, and branches from the very ground. Its leaves, which are narrow, and shaped like a short straight swordblade, have near the top of the stem a remarkably broad sheathing base. At the extremities of the branches appear round balls of flowers, some of which (*b. b. fig.* 1.), are bright green, and others (*a. a.*) bright yellow; the latter being the most numerous, and placed above the others. The yellow heads consist of stamen-bearing, or barren flowers, and the green heads of pistil-bearing, or fertile florets. What happens in this case, occurs also in all instances in which the stamens are separated from the pistils in different flowers on the same plant; we invariably find that the stamens are placed on the uppermost parts of the branch above the pistils, an arrangement which is no doubt provided, to facilitate the scattering of their pollen upon the stigmas. If they were placed below the pistils, it would be much more difficult for the pollen to reach the stigma, and consequently, the great end of the creation of the stamens would be almost frustrated. We find, however, that every thing is foreseen, and provided for by Providence, with the same care in these little plants, as in the most exalted and perfect of the works of nature ; and that even so apparently useless and insignificant a weed as the Bur-reed, contains the most convincing evidence of the worthlessness of the opinions of those who, denying the existence of the Deity, would have the world believe that living things are the mere result of a fortuitous concourse of atoms, attracting and repelling each other with different degrees of force.

If we open one of the yellow balls, we find it consists of a great number of separate flowers, each of which has a calyx of three long-stalked jagged sepals (*fig.* 2. *a. a. a.*), and six wedge-shaped anthers, which are heavier than the little slender filaments can well support.

In the structure of their calyx, the flowers of the green heads do not materially differ from the latter (*fig.* 3.); the sepals are three in number, but broader and shorter, rolled round the pistil, and seated close

upon the receptacle, without any stalk. The pistil is
an oval body, terminated by a deeply lobed stigma
(*fig.* 3.), and having in its ovary but one cell, from
the summit of which hangs a single ovule. The
fruit contains one seed (*fig.* 6.), consisting of a mass
of albumen (*fig.* 7.), at the further end of which lies
a minute embryo (*fig.* 8.).

It is obvious that this is an organization much more
imperfect than that of the Monocotyledonous orders
we have before examined; in room of an evident
calyx and corolla, we have nothing but three scales,
there is no trace of the number three in the ovary,
and the stamens and pistils are separated from each
other. In the Bullrush itself, the imperfection is yet
greater; even the scales, which in the Bur-reed represent the calyx, are wanting, and nothing appears
in their room but a quantity of delicate black hairs.

Nearly allied to the *Bullrushes*, like them useless
to man, as far as we know, and equally inhabiting
wet and spongy soils, are some little inconspicuous
plants, called by the learned *Triglochin*, and by
others *Arrow-grasses.* These, and some allied genera,
form an order called the *Arrow-grass tribe* (Juncagi-
neæ, Plate XXI. 2.), in which incompleteness of
structure exists in a less degree than in the Bullrushes, for they have a calyx and corolla, and the
usual number of stamens; but the former are more
like scales than sepals and petals, and the ovary, although it has the usual ternary structure of Monocotyledons, is nevertheless in an imperfect state.

Marsh Arrow-grass (Triglochin palustre), is a little

inconspicuous plant, which does not grow above
eight or nine inches high, and, both its leaves and
stems being remarkably slender, is easily overlooked.
Its curious structure will, however, amply repay
the trouble of finding and examining it. It is
common enough in wet meadows, or in other moist
places, where the ground is covered with a sward of
grass, from which you will never distinguish it, without employing all your power of observation. It is
said to send out scaly runners, which end in little
knobs, shaped like a scorpion's tail. Its very narrow
succulent leaves are of just the length and colour of
those of many grasses, but they are taper, and not
flat. When about to flower, it throws up from the
midst of its leaves a little slender undivided stem
(*fig.* 1.), which bears at its upper end a good many
green flowers, loosely arranged at some distance
from each other. Each of these flowers is constructed
thus : On the outside are three concave blunt scales
(*fig.* 2. *a. a. a.*), which constitute the calyx ; in the
cavity of each of these there lies a roundish anther
(*fig.* 4. *b.*), the face of which is next the scale,
and consequently in the most unfavourable position
possible for discharging its pollen upon the stigma.
Within the calyx are three other scales, which are
pressed close to the ovary, and which in like manner
contain three anthers in their hollows ; these scales
are the petals; so that an *Arrow-grass* has three
sepals, three petals, and six stamens. Its ovary is a
long three-cornered body, having no style, and for
a stigma nothing but a tuft of little hairs (*fig.* 2.

& 3. *b.*). It may be separated with care into three carpels, at the bottom of each of which lies a single erect ovule, a great deal smaller than the cavity in which it is placed (*fig.* 3.). The fruit is a narrow three-cornered dry case, which divides into three slender parts, each of which is partly filled with a taper seed (*fig.* 7.). In almost all Monocotyledonous plants, a considerable quantity of albumen is provided for the nutriment of the young embryo, but the *Arrow-grass tribe* is one of those which form an exception to this rule. The embryo of these plants lies immediately below the seed-coat, and consists of a cotyledon (*fig.* 8. *c.*), having near its base a small slit (*a.*) through which the young stem is protruded when the plant begins to grow.

It is evident that this kind of structure is so unlike what we find in the Bullrushes, as to require no further explanation. What does require elucidation is the reason of the singular arrangement of the anthers, which I have just described; it does not at all appear for what cause they are so carefully embodied in two hollows of the calyx and corolla, nor indeed how, under such circumstances, their pollen is ever to reach the stigma. I must confess my inability to explain the matter; you cannot do better than reflect upon it, until you hit upon some solution of the mystery.

Far more simple than the previous tribes, is that to which the *Duckweed* (Lemna) gives its name. In the former, whatever deficiency there might be in the parts of fructification, there was at least a stem and leaves. But in Duckweed there is nothing but a fleshy floating green body, which looks like a green scale, and which is in reality a compound of both root and stem. Most people fancy that Duckweed never flowers ; and many are they who have watched it all their lives without succeeding in discovering its blossoms. If, however, you will fix your eyes attentively upon a mass of it, on a still sunshiny day in the months of June or July, you will probably dis. cover exceedingly minute straw-coloured specks here and there on the edges of the plants ; they have a sparkling appearance, and notwithstanding their minuteness readily catch the eye. These are the anthers, and they being found, you have only to carry home the plants, and place them under a microscope when all the secrets of their flowering stand revealed. Where the anthers have caught the eye, will be seen a narrow slit, out of which they peep; if you widen this slit (Plate XXI. 3. *fig.* 2.) with your dissecting instruments, you will be able to extract the blossom entire (*fig.* 3.); and you will have before your eyes the simplest of all known flowers, as Duckweed itself is the simplest of all known flowering plants. The flower consists of a transparent membranous bag, shaped like a water caraffe, and split on one side; within it are two stamens, and one ovary with a style and simple stigma. The fruit (*fig.* 4.) contains but one cell, in which are one or more seeds (*fig.* 5.); its shell is a thin cellular integument.

Such are the simple means that Duckweed possesses of propagating itself; means, however, which appear to be abundantly sufficient, if we are to judge from the immense quantities which rise every year to the surface of our ponds. In Europe we have no other plant belonging to the same natural order; but in tropical countries its place is occupied by a plant called *Pistia*, which is a sort of gigantic Duckweed, with broad lobed leaves like some Lichens, and a more highly organized flower.

With these examples of imperfect Monocotyledonous plants, I must dismiss that part of the subject; in my next letter I shall introduce you to the more interesting and important tribe which furnishes us with bread.

EXPLANATION OF PLATE XXI.

I. THE BULL-RUSH TRIBE.—1. A shoot of *Branching Bur-reed* (Sparganium ramosum); *a a* heads of barren flowers; *b b* heads of fertile flowers.—2. A barren flower; *a a a* sepals.—3. A fertile flower. —4. An anther seen from the side.—5. The same viewed from the edge. —6. A seed.—7. The same divided perpendicularly, with the embryo in one end of the albumen.—8. The embryo extracted.

II. THE ARROW-GRASS TRIBE.—1. A flower-spike of *Marsh Arrow-grass* (Triglochin palustre).—2. A flower; *a a a* sepals; *b* stigma.— 3. A perpendicular section of one of the cells of the ovary; *b* stigma. —4. A scale of the calyx *a*; with the anther, *b*, lying in it.—5. An anther seen on the side by which it discharges its pollen.—6. A ripe fruit.—7. A perpendicular section of one of its cells, shewing the manner in which the seed lies in it.—8. An embryo; *a* the slit through which the stem is finally protruded; *b* the radicle; *c* the cotyledon.

III. THE DUCKWEED TRIBE.—1. Plants of *furrowed Duckweed* (Lemna trisulca) floating on water.—2. A plant magnified, and in flower at *a*.—3. A flower extracted from the slit in which it laid; *a* the membranous bag.—4. A fruit.—5. A seed.

LETTER XXII.

GLUMACEOUS PLANTS—THE GRASS TRIBE—THE SEDGE TRIBE.

(Plate XXII.)

For the natural orders forming the two different degrees of developement we have now considered, there is no collective name; but for the third we have one in such general use, that we must not pass it by in silence.

There are certain plants which are called GLUMA-CEOUS, because in place of calyx and corolla they have nothing but green or brown scales, named *glumes*, arranged alternately round a common centre. The most remarkable of such plants are Grasses and Sedges, to the former of which I propose first to call your attention.

The most common and the most useful of all the natural orders of plants is beyond doubt the *Grass tribe*; wherever we cast our eyes, we are sure to see blades of grass springing up, if any vegetation at all can exist; it is they that form that universal verdure which gives the northern parts of the world their peculiar charm, and which alone may console the inhabitants for the want of those other advantages

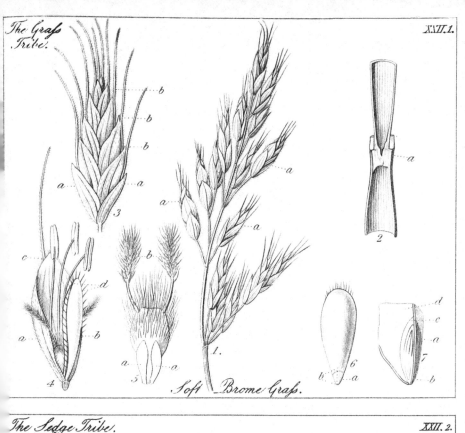

Soft Brome Grass.

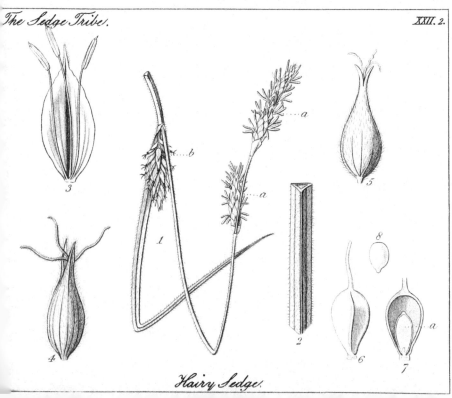

Hairy Sedge.

which a brighter sun and more cloudless sky are capable of supplying. Bread, is exclusively prepared from the flour or albumen of the seeds of various kinds of corn, chiefly from wheat, and the richness of pastures depends essentially upon the species of Grasses that inhabit them. If it were not for the creeping subterranean stems of maritime Grasses, which can vegetate amidst dry and drifting sand, the banks which man heaps up as a barrier against the ocean would be blown away in the first hurricane; but the *Sea-reeds* (Ammophila), *Lyme-grasses* (Elymus), *Wheat-grasses* (Triticum), and others, vegetate rapidly on such embankments, and, piercing the soil in every direction with their tough underground stems, presently form an entangled web of living matter, which is spontaneously renewed as fast as it is destroyed, and which offers a resistance to the storm which is rarely overcome. Finally, the Bamboo alone is capable of supplying all the wants of savage man; with its lightest shoots he makes his arrows, thin strips of the wood form bow-strings, and from the larger stems he fabricates a bow; a long and slender shoot affords him a lance shaft, and he finds its hardened point a natural head for the weapon. With the larger stems he builds the walls and roof of his house; its leaves afford him an impenetrable thatch; split into narrow slips it gives him the material for weaving his floor mats, and other articles of domestic convenience; its fibre furnishes him with twine, and its leaves provide him with paper, when he becomes sensible of the utility of such a material.

Would he commit himself to the waves, the stems form the hull of his boat, which a few skins stretched over it render water tight ; they also give him masts ; and thin slips of wood become cordage or are woven into sails.

In tropical countries Grasses are far more gigantic than in England ; we usually see them at their largest, two or three feet high, when in flower in the hay fields ; and the reeds that in marshes or ponds gain a stature of seven or eight feet, are probably the noblest specimens of the tribe with which you are acquainted. But in equinoctial regions, where the air is damper, and the sun far more powerful and brilliant than with us, Grasses acquire such surprising dimensions as to rival Palms themselves in majesty of appearance. In Brazil we are told that the hay will grow seven or eight feet high, the *Sugar-cane plant* (Saccharum) averages 20 feet, and *Bamboos*, with their light imperishable stems, lance up into the air to the height of 30 or 40 feet. It is in such regions alone that we can really behold the perfection of the Grass tribe.

Let us, however, be contented with examining one of our own species for a knowledge of the structure of this singular tribe ; we shall find every peculiarity of structure that tropical species afford, and we can easily imagine all that difference of size is likely to produce.

One of the commonest of British Grasses is the *Soft Brome-grass* (Bromus mollis, Plate XXII. 1.) ; a plant which you will be almost certain to meet with in the first piece of waste ground. It is an annual with an erect stem, about two or three feet high, and covered all over with soft hairs. The stem requires more than mere external examination. Strip it of its leaves, so as to lay bare all its surface ; and you will find it hard and thickened at every joint where a leaf has been torn. Split it, and instead of the solid centre of other plants, you will see that it is hollow, and consists of nothing but a cylindrical shell (*fig.* 2.); at the joints, however, the sides of the cylinder meet, and form a firm partition which completely separates one part of the stem from the other. It is this structure that renders the Bamboo so useful for forming cases to hold rolls of paper ; in India they cut a truncheon of a stout Bamboo, and scrape away all the partitions except one ; it then becomes a cylinder open at one end, where the partition is destroyed, and closed at the other end. A short piece of a Bamboo of the same diameter, having a complete partition at one end, is then formed into a lid by scraping away its inside, and a capital case is produced without farther trouble. We sometimes see such cases, as much as two feet in circumference.

The leaves of Brome-grass are hairy, narrow and sharp pointed at their lower end, which, notwithstanding its breadth, is considered to be their stalk ; it rolls round the stem, forming a kind of sheath, which sometimes is not very easily unrolled. At the upper end of the sheath you may remark a thin white membrane, such as you have nowhere met with before ; Botanists call such a membrane a LIGULA.

Thus far then we have two peculiarities in Grasses;
their *hollow round stems, with partitions* at the joints,
and their *ligulate leaves.*

The flowers are still more unlike what you have
before seen. At the top of the stem of the Brome-
grass, a number of slender branches appear, turned
chiefly towards one side, and by their weight giving a somewhat nodding appearance to the parts
they bear (*fig.* 1.); those parts (*a. a. a.*) are oblong
green bodies, apparently composed of scales after the
manner of a leaf bud; in reality they are little collections of flowers, whence they are named *spikelets*
(or spiculæ or locustæ, as they say in Botanical Latin).

Each spikelet is constructed as follows. Firstly,
at its base (*fig.* 3. *a. a.*) are two green scales, each of
which has about five ribs; these are the GLUMES
strictly speaking; there is no trace of either pistils or
stamens in their bosom, on the contrary, they are always found to be perfectly empty. When the glumes
are removed you come to some other parts, which at
first sight look like glumes; but on a more careful
inspection you will remark that they are composed of
more scales than one, have a stiff bristle at their
back, and contain some stamens, &c. in their bosom.
These are called FLORETS; in the Brome-grass there
is about ten of them placed one above the other in
two opposite rows (*fig.* 3. *b.*).

Each floret consists of two scales called PALEÆ
(*fig.* 4. *a.* & *b.*); of which the more external (*a.*) is the
larger, is covered all over the outside with soft hairs,
and bears at its back a little below the end (*c.*) a stiff

bristle, called the BEARD, or AWN (arista); the beard
is in reality the midrib of the palea, partially separated and lengthened out. The inner palea (b.)
originates from above and within the base of the
outer, is much smaller and more membranous, has
its edges abruptly doubled inwards, and bears a row
of stiff bristles on the angles (d.) formed by the
doubling. These two are the lower or outer, and
upper or inner paleæ.

Next the paleæ come, on the side next the outer
palea, two exceedingly small scales (*fig. 5. a. a.*),
which are much shorter than the ovary; they are
called HYPOGYNOUS SCALES, and are supposed to be
the rudiments of a calyx or corolla.

From the base of the ovary arise three stamens
(*fig. 4.*), whose filaments are white, and so weak and
slender, that the long narrow anthers hang in a state
of oscillation, in consequence of the inability of the
filaments to support them.

The ovary is a wedge-shaped body, apparently
consisting of nothing but pulp, and crowned by a tuft
of long hairs (*fig. 5.*); two styles, bearing singular
brush-like stigmas, spring from its summit.

In this instance you have all the parts that are
usually present in Grasses; and you cannot avoid
remarking how widely different the whole organization is from any thing you have witnessed in other
plants. The structure of the fruit is not less peculiar.

I have said that the ovary seems as if it were composed of nothing but pulp; it does, however, consist

of an ovule, and of a shell that includes it, but both
are so soft that they grow together, and cannot be
distinguished. Immediately after the styles wither,
the ovary swells, gradually loses its softness, and at
last when ripe is nearly bald, having gained a sallow
appearance, and become longer and thinner. At the
period of maturity (*fig.* 6.) there is still no means of
separating the shell of the fruit from the skin of the
seed, so completely are they grown together; the
fruit looks therefore so like a seed, that it is no
wonder it should popularly be called so; it is better,
however, to designate it a GRAIN. If you crush the
ripe grain you will find its contents of a hardish
horny consistence, but easily reduced into the state
of flour; from what you have seen in other instances
you will easily recognize this for albumen. It is
possible, that you may search in vain for an embryo,
amidst all this flour; and I dare say, if I do not tell
you how to look for it, you will waste a great deal of
time in finding it, even if you should recognize it
when found. Follow me attentively, and I shall
easily relieve you of this difficulty. The ripe grain
is much narrower at one end than the other, and
more convex on one side than the other; turn the
grain on its flat side, so that the convexity is uppermost, and then carry your eye to the narrowest end;
there you will espy a minute oval depression (*fig.*6.*a.*);
if you carefully lift up the skin at this part, you will
detect the embryo lying snugly half buried in albumen. It will appear like a greenish-yellow plano-
convex oval body, in which you can discern no marks

of organization. But if you will divide it perpendicularly with a sharp knife, you will then be able to see that it has a most complete and highly developed structure. You will find (*fig.* 7.) that it consists of a thickish scale (*c.*) upon which lies a little conical body (*a.*), composed of several minute sheaths fitted one over the other ; the scale is the cotyledon, and the conical body the plumule or young stem. At the lower end of the embryo may also be made out a sort of sheath lying within its extreme point (*b.*) ; it is the rudiment of the root.

When the embryo first begins to grow, the cotyledon (*c.*) swells a little and attaches itself firmly to the albumen by the whole of its highly absorbent surface : the albumen at the same time softening and becoming partially dissolved by the moisture it has taken up from the soil ; by this means the nutritive matter of the albumen is conveyed into the cotyledon as quickly as it is formed. The food thus poured into the cotyledon by thousands of invisible mouths, causes it to swell and all its parts to lengthen. The radicle (*b.*) is pushed downwards into the soil on the one hand, and on the other the plumule rises upwards into the air ; both these parts are abundantly supplied with the materials of growth by the cotyledon, until the roots have established themselves in the soil, and are able to pump up food for themselves, and for the nascent stem. By the time this happens the cotyledon has shrivelled up, the albumen is exhausted of its nutriment, and all these temporary parts cease to exist.

Such is the provision that nature has contrived to ensure the perpetuation of Grasses ; for there is but little variation from this arrangement throughout the species of the tribe.

Here let us pause for an instant to admire the beautiful adaptation of all the parts to the functions they have to perform. Grasses spring up with rapidity as soon as the earliest rains have fallen upon the dry ground where their grains have been deposited ; it is necessary that they should do so in order that, however imperfect the supply of rain may be, the earth may at all times be clothed with verdure. To ensure this, Providence has given them a young stem which is almost formed in the very seed, and which is ready upon the slightest stimulus to spring forth into life ; but if the young stem were to sprout with much rapidity, the roots would be unable to supply it fast enough with food, and it would presently wither and die, unless some special means were provided of meeting this difficulty ; accordingly, a great abundance of albumen is stored up, as a certain supply of food till the roots can themselves obtain it from the earth. The supply of albumen would, however, be useless unless some means existed of conveying it with rapidity to the plumule, and accordingly we find the broad thin cotyledon, a highly absorbent body, placed with its whole surface applied to the albumen, and ready to transfer the nutritive fluid to the plumule as quickly as the former can be formed.

Grasses are so numerous and so very simple in

their structure that you may well believe there is some little difficulty in distinguishing them into genera; especially as their parts are so small. Although it is no part of my plan to teach you much of the distinctions of genera, that being left you to acquire from systematic works, when the difficulties of this introduction are mastered; yet I may as well explain how a few of the very commonest genera are known from each other.

What has already been said of the Brome-grass explains its characters. Very nearly related to it are *Fescues* (Festuca), the species of which, whether highlanders or lowlanders, are so much and advantageously employed as pasture Grasses; they differ from the Brome-grasses only in having their beard proceeding from the very point of the palea, instead of from below its point. Like both these are the *Meadow-grasses* (Poa); but they have no beards at all, and are usually much smaller : the little annual Grass which grows every where, and flowers at all seasons of the year, lying almost prostrate upon the ground is a kind of Poa (P. annua). *Quaking-grasses* (Briza) are Poas with their paleæ inflated. All the above have a loose inflorescence, and several florets in each spikelet.

Others, having also a branched inflorescence, are known by each spikelet containing but one floret; such as *Bent-grasses* (Agrostis), with thin delicate silken panicles and wiry stems; they have often only one palea instead of two, or if they have two the upper one is very minute. To Grasses with a similar structure also belong the *Feather-grass* (Stipa), with its long and plumed beards, and *Catstail-grasses* (Phleum), with two equal sharp-pointed glumes; *Canary-grass* (Phalaris), too, with whose grains your Canary birds are fed, and whose glumes have each a flat keel like a little boat, and *Foxtail-grasses* (Alopecurus), which differ from the *Catstails* in having a beard to their single palea. In the *Catstails, Canary-grasses,* and *Foxtails* you will not recognize at first a branched inflorescence, for two of them derive their names from the compact appearance their flowers present. But if you separate the flowers gently, you will find that they are in reality seated upon little branches, which are pressed so closely together that you do not see them.

Another group of Grasses has the spikelets really seated close to the stem; as for instance *Wheat, Barley,* and *Rye;* while a fourth kind has the stamens in one kind of flower and the pistils in another: of this kind is Maize, or Indian Corn. In that plant the barren flowers are loose yellow branches, growing at the top of the stem, and covered with anthers, while the fertile flowers are hidden among the lower leaves, and are only discovered by their long shining styles, which hang down in tufts like silken tassels.

The other natural order of Glumaceous Monocotyledons, which I propose to mention to you is that which comprehends *Sedges,* after which it is named. In its general appearance it resembles Grasses; but it is known by its stems being solid, not hollow (Plate XXII. 2. *fig.* 2.), and by its leaf-stalks, when they roll round the stem, growing together by their edges into a perfect sheath. These distinctions are the more important from being accompanied by others in the parts of fructification, and also by an absence of all those useful properties for which Grasses are remarkable. The most common genus of the whole order is the *Sedge* (Carex) ; a species frequent on wet commons, called " the hairy" (C. hirta), will supply us with an illustration. Its solid triangular stem (*fig* 2.), and its hairy leaves agree with what has just been mentioned. Its flowers are arranged in heads, and are of two kinds ; one sort occupying the upper end of the stem (*fig*. 1. *a. a.*) consists of barren flowers only ; each flower (*fig*. 2.) has an oval brownish membranous scale and three stamens, and the heads are composed of nothing but such flowers. The other kind of head (*fig*. 1. *b.*) appears a little below the others, is green, and consists of fertile flowers only. Like the barren heads this is chiefly composed of imbricated scales, but in place of the stamens you find (*fig*. 4.) a hairy bottle-shaped body, split at the end into two lobes from between which three stigmas project. Open the bottle and you will discover that the stigmas are connected with a single style, springing from the top of a three-cornered ovary. This is all that the fertile flowers consists of ; the bottle is formed by two scales which are placed opposite each other and grow together at their edges, and is a mark of the genus Carex ; most others of the Sedge tribe are without it, and contain nothing but a naked pistil. When the fruit

(*fig.* 6.) of the Carex is ripe, it is still enclosed in the
bottle (*fig.* 5.), but it has become a hard 3-cornered
brown nut, with a thickish shell, and one seed standing
erect in its cavity (*fig.* 7. *a.*).   As for its embryo it
is totally different from that of Grasses, it neither lies
on the outside of the albumen, nor is it shaped at all
like it; it is a minute roundish undivided body
(*fig.* 8.), which is buried in the lower end of the
albumen.   It really would seem as if the small comparative utility of these plants was indicated by the
little care that nature has taken to ensure the growth
of their seeds by extraordinary precautions.

It is to the Sedge tribe that most of those plants
belong, which are popularly considered Rushes, and
which afford the materials of the manufacture of
candles, mats, and chairs; the *Club-rush* in particular (Scirpus lacustris), which sometimes grows as
much as nine feet high, is the species that is collected for such purposes.   They are chiefly found on
wet commons, or in marshy, or swampy places.   The
most remarkable of the wild kinds is the *Cotton-
grass* (Eriophorum), the long silky white hairs of
whose fruit look exactly like tufts of cotton blowing
about in the wind.

Here ends my explanation of the organization of
those plants which, because they are increased by the
action of an apparatus consisting of calyx, corolla,
stamens, and pistil, and forming what we call a
flower, are called FLOWERING.

The remainder of the Vegetable Kingdom consists
of species wholly destitute of flowers, and increased by

organs totally different in their nature from fruit and seeds. For an explanation of the characters of these, I must refer you to the succeeding letter.

EXPLANATION OF PLATE XXII.

I. THE GRASS TRIBE.—1. A piece of the inflorescence of *Soft Brome-grass* (Bromus mollis); *a a* spikelets.—2. A perpendicular section of a portion of the stem, shewing the partition at *a.*—3. A spikelet : *a a* glumes ; *b b b* florets.—4. A floret half open ; *a* the lower palea ; *b* the upper palea ; *c* the beard ; *d* the angles of the upper palea covered with stiff hairs.—5. A pistil ; *a a* hypogynous scales; *b* stigmas.—6 A ripe grain ; *a* the place where the embryo lies ; *b* the piece which is cut out, and seen magnified at 7, where *a* is the plumule ; *b* the radicle ; *c* the cotyledon, and *d* the albumen.

II. THE SEDGE TRIBE.—1. A portion of the upper end of the *Hairy Sedge* (Carex hirta); *a a* barren flower heads ; *b* fertile flower head.—2. A portion of the solid stem.—3. A barren floret seen from the inside, with its three stamens.—4. A fertile floret.—5. A ripe fruit, as seen when enclosed in the bottle.—6. The true fruit taken out of the bottle.—7. A perpendicular section of the same, shewing its erect seed *a.* —8. An embryo.

LETTER XXIII.

CELLULAR, FLOWERLESS, OR CRYPTOGAMIC PLANTS—
THE FERN TRIBE—THE CLUB-MOSS TRIBE.

(Plate XXIII.)

WE have now arrived at the frontiers of the third great province into which the Vegetable Kingdom is divided; we are about to visit races far less perfectly organized than those of the two other provinces; and we shall no longer be delighted by beautiful form, or astonished at stupendous size, or interested by the many useful purposes to which the species are applied. We here lose sight of all traces of flowers, and of the singular parts that belong to them; in their room we find a different mode of propagation, and along with it, a plan of organization which is always far more simple than in Flowering plants, and which gradually diminishes in complexity, till at last the component parts of the vegetable texture are separated, and nothing remains bnt little bladders, which are hardly to be distinguished from the simplest forms of animals.

Plants of this sort are called FLOWERLESS; they are also named *Cryptogamic*, because the parts by which they are by some thought to be increased are hidden from view; and they have, in addition,

The Fern Tribe. *XXIII. 1.*

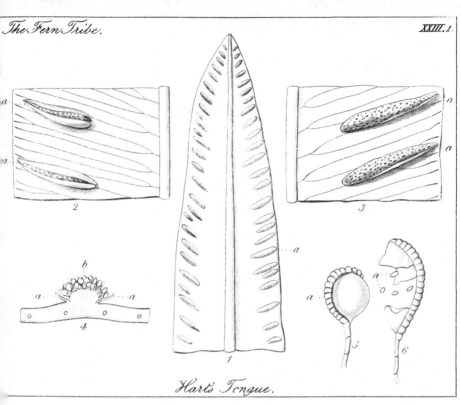

Harts Tongue.

The Club-moss Tribe. *XXIII. 2.*

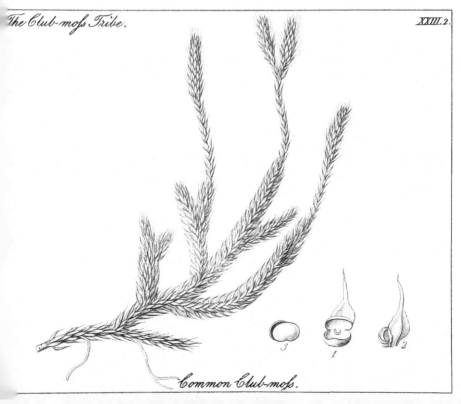

Common Club-moss.

the designation of CELLULAR plants, because their
stems rarely contain vessels, but usually consist of
cellular substance only. You, perhaps, do not know
that, in nearly all plants which bear flowers, there
are present those curious air-pipes, called spiral vessels, of which I long ago gave you an explanation
(p. 38.); such, however, is the fact. On the other hand,
in these Flowerless plants, spiral vessels are universally absent; some of the most highly organized
tribes, such as those which form the subject of the
present letter, have a particular kind of vessel in lieu
of the spiral; but all the other tribes are destitute of
vessels of any kind.

Ferns, which are the most completely organized of
Flowerless plants, and which approach nearer than
any others to the Flowering tribes, are those to which
I would first call your attention. In the northern parts
of the world, they are green leafy productions, which
die down to the ground every year; and they are
seldom more than two or three feet high; one of the
larger kinds is the *Brake* (Pteris), which is so much
esteemed as covert for game. But, in tropical countries, many of them far surpass these pigmy dimensions; they acquire real trunks, resembling those
of Palms, and often rise to as high as forty or fifty
feet without a leaf; even a more considerable stature is spoken of by travellers. At all times they
are graceful objects, from the slender wiry stems
on which they bear their leaves, which wave in
the breeze like plumes of feathers, and from the
multitude of leaflets into which they are cut with

the most exquisite regularity. But the Tree Ferns of the
Tropics are said to be most superb objects, combining
the grace and agreeable colour of their European
kindred with the majestic aspect of the Palms.

It is usual to call the leaves of Ferns by the name
of FROND, as if their leaves were not analogous to
those of other plants. But I see no use in continuing
this old fashioned word, which was coined at a time
when the leaf of a Fern was thought to be a sort of
compound of a branch and a leaf. It is much better,
on every account, to call it by the name that the
same part bears in other plants.

For the purpose of studying the organization of
Ferns, I recommend you to take a leaf of *Hart's-tongue*
(Scolopendrium officinarum, Plate XXIII. 1.), a
plant which is common on most damp and shaded
banks, and within old open wells, the mouths of which
are often almost choked up by it. All that you will
find of the plant, is a brown scaly rootstock, from
which grow a number of handsome lance-shaped
leaves, of a deep green colour, placed upon a shining
ebony-black stalk. If the leaves are newly formed,
you will, by holding them up against the light, readily
see their veins, which are dissimilar to those of all
other plants. They neither resemble Monocotyledons,
nor Dicotyledons; are neither netted nor parallel;
but have a simply forked structure. You will remark, that, although now and then, one vein may
be found running straight from the midrib nearly up
to the margin, without dividing, yet, that the principal part fork very soon after the vein has left the

midrib, and that, sometimes, one of the branches forks again. This kind of vein is peculiar to Ferns, and will enable you, at all times, to recognise them whether their reproductive parts are present or not.

After the leaf has been growing some little time, you may remark a number of narrow pale bands appearing at pretty equal intervals upon some of the veins, and following their direction (*fig.* 1. *a.*). Presently afterwards the whole of the skin of the leaf, where these bands are, separates from the green part below it : in course of time, something swells and raises up the skin, till at last it bursts through it, separating the skin into two equal parts, one edge of which remains adhering to the leaf (*fig.* 2. *a. a.*). At this period the cause of the swelling is discovered ; it consists in a multitude of brown seed-like grains that are crowded together very closely, and form a brown ridge (*fig.* 3. *a. a.*). Botanists call the skin which separates from the leaf the INDUSIUM, the ridge SORUS, and the seed-like grains THECÆ. In order to gain a distinct view of all these parts, you should cut through the leaf across a sorus, just after the indusium has burst ; and then the edges of the indusium will be distinctly visible (*fig.* 4. *a. a.*), with the ridge-like receptacle of the thecæ rising up between them (*fig.* 4. *b.*).

The only means of propagating itself, which the Hart's-tongue possesses, resides in the thecæ. It has no calyx, corolla, stamens, or pistillum, and consequently neither fruit nor seed ; nevertheless it can perpetuate its kind with the same certainty as the most perfect plant. The theca (*fig.* 5.) is not a seed,

nor is it a body whose functions are of a nature similar
to those of the seed ; you require a pretty good microscope to examine it correctly, but with such an instru-
ment you will make it out to be a roundish compressed body, seated on a jointed stalk, which runs
up one side of the theca (*fig. 5. a.*).   Upon examining a good many of the thecæ, you will no doubt
remark some of them burst open (*fig. 6.*) ; and then
you will find that they are hollow bodies, containing
a quantity of extremely minute oval grains (*fig. 6. a.*),
called SPORES, by Botanists.   It is in the spores that
the power of increase resides ; every one of them
will form a new plant, and consequently they are
analogous to seeds; but as they do not result from
the action of pollen upon a stigma, they are not real
seeds, but only the representatives of those organs
amongst Flowerless plants.

How simple is all this; how different from every
thing we have seen in other plants ! and yet no
doubt as perfectly adapted to the multiplication
of Ferns as any more complete contrivance.   How
prodigious too is the power that these plants possess
of disseminating themselves! Hart's-tongue, owing
to its small size, is one of those in which the power
resides only in a small degree ; and yet a little com-
putation will shew even its means to be prodigious.
Each of its sori consists of from 3000 to 6000 thecæ ;
let us take 4500 as the average number.   Then each
leaf bears about 80 sori ; which makes 360,000 thecæ
per leaf; the thecæ themselves contain about 50
spores ; so that a single leaf of Hart's-tongue may

give birth to no fewer than eighteen millions of young plants.

The form and situation of the sori is not, in other genera, the same as in the Hart's-tongue; on the contrary it is upon differences in those respects that the genera have been established. For example in *Shield-ferns* (Aspidium) the sori are round and covered with a kidney-shaped indusium; in *Polypody* (Polypodium) they are round and have no indusium; and in the graceful *Maiden-hair-ferns* (Adiantum) they are oblong bodies arising from the edges of the leaf. The most curious arrangement of their parts is in the *Brake* itself (Pteris); no matter at what time of the year you examine the leaves of that plant you will probably discover no trace of sori, and yet it would be difficult to find a Brake-leaf in the autumn which does not abound with them. The truth is that in this plant they occupy so singular a position that one could almost be tempted to believe that they were designedly hidden where none but the curious Botanist should find them. Look attentively at the under side of the leaves: you will remark the margin to be turned in and thickened, like the hem of a lady's gown in which a cord is run; there lurk the thecæ you are in search of. With the point of a knife lift up gently the edge of the leaf, and you will at once discover a ridge of thecæ running all round it; in this instance the margin of the leaf acts the part of indusium.

Another singular form of Ferns is that in which the whole of the segments of a leaf are contracted and

curled up round the thecæ, so as to lose entirely the
natural appearance, and to resemble a sort of inflorescence. A striking instance of this is not uncommon in bogs, in the form of a plant called the *Os-
mund-royal*, or *Flowering Fern* (Osmunda regalis); a
minute species found in woods, and called *Adder's
Tongue* (Ophioglossum), because of its narrow inflorescence, is another British example.

Such is the first and highest degree in the scale of
organization among Flowerless plants. Possessing a
system of vessels, frequently attaining a considerable
size, having leaves intersected by veins, and having
their surface provided with breathing pores, Ferns
may be considered to differ from Flowering plants in
little except in the manner in which they are propagated, and in the organs assigned them by nature for
that purpose. Next to them is arranged a small tribe
also possessing a system of vessels in the stem, and
breathing pores on the surface, but destitute of veins,
and having a remarkably different mode of repro-
duction; you will find, indeed, that there is this
great peculiarity in Flowerless plants, independently
of all others, that no two tribes agree exactly in the
nature of their organs of propagation. While in
Flowering plants one tribe is distinguished from another by slight variations in the form, or number, or
proportions of a few organs that they all possess in
common, you will find among Flowerless plants, on
the contrary, that every tribe has an independent and
peculiar provision of its own for the perpetuation of
the species.

This is the case in the *Club-moss tribe* (Lycopo-
diaceæ, Plate XXIII. 2.), to which I have alluded.
Club-mosses, in some parts of England called also
Snake-mosses, are humble plants which grow on moors
or heaths, or half-drained bogs, over which their
scaly stems creep and interweave. There are no veins
in their leaves, which are for the most part narrow,
and taper-pointed. When about to reproduce themselves, they emit from the ends of their branches,
which are usually forked like the veins in a Fern-
leaf, a slender shoot of a paler colour than the remainder, and terminated by a yellowish thickened
oblong, or club-shaped head. Among the hair-
pointed leaves of the head lie, one in the bosom of
every leaf, pale yellow cases, opening by two or three
valves (*fig.* 1.), and containing either a fine powdery
substance, or a few large grains or spores. These are
all the means such plants have of propagating themselves; and it is uncertain what the exact difference
is in the purposes to which the powder and the
spores are severally destined. The latter, no doubt,
grow like seeds, but it is not quite certain that the
powder grows also; there are those who say they
have seen the powder grow, but their observations
require to be repeated.

Such plants seem to occupy an intermediate place
between Ferns and Mosses, to the latter of which my
next letter will refer. I will not detain you about them
further than by remarking, that although they are now
seldom more than three or four feet long, and are generally much smaller, it is probable that either similar

T

plants, or races very closely allied to them, grew in ancient days, long before the creation of man, to a size far beyond anything that the present order of things comprehends; this is, however, a geological matter, with which we had better not interfere.

EXPLANATION OF PLATE XXIII.

I. THE FERN TRIBE.—1. The upper end of a leaf of *Hart's-tongue* (Scolopendrium officinarum), seen from the under side; *a* sori.—2. A portion of a leaf magnified, to show the veins and the structure of the sori, *a a*, more distinctly.—3. A similar portion, further advanced; *a a* sori.—4. A section of a portion of a leaf across the sorus; *a a* the two valves of the indusium; *b* the sorus itself, covered with thecæ.—5. A theca; *a* its elastic back.—6. Another theca burst, and scattering its sporules, *a*.

II. THE CLUB-MOSS TRIBE.—A plant of *common Club-moss* (Lycopodium clavatum).—1. A theca, with its two valves open, and the leaf out of whose bosom it grows.—2. The same leaf seen from behind. —3. A theca, as seen when closed.

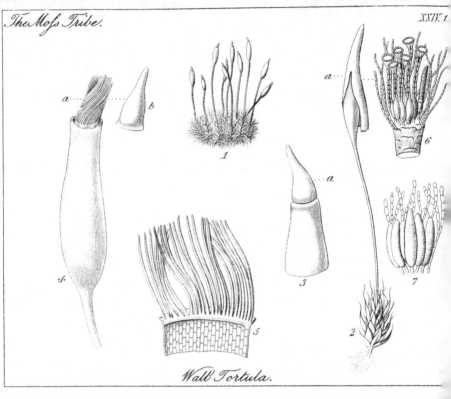

Wall Tortula.

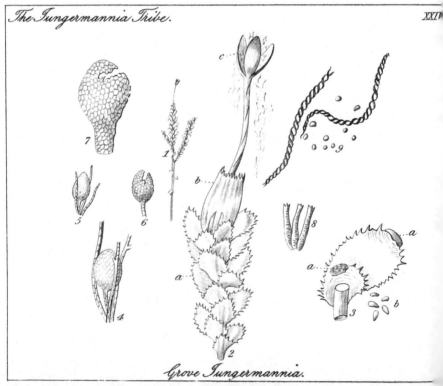

Grove Jungermannia.

LETTER XXIV.

Plate XXIV.

STILL lower in the scale of creation, than Ferns and Club-mosses are the true *Mosses;* plants destitute of all traces of vessels and of breathing pores; with no true veins in their leaves, and so pigmy in growth, that the most gigantic of them hardly equal the smallest of the Club-mosses.

Up to the present moment, a microscope has rarely been necessary in our studies; whenever I have recommended you to employ it, the subject would usually have admitted of your dispensing with its aid, if you had pleased. But from this time forward it must be constantly in your hand, and every observation must be made with it. You will, however, find abundance of most curious and interesting results, to indemnify you for the trouble it will give you.

Mosses are among the smallest of plants with true leaves; they are often so minute that the whole specimen, leaves, stem, fruit, and all, would escape the eye, if they did not grow in patches; and they never, in the largest kinds, exceed the height of a few inches. Nevertheless, they are organized in a man-

ner far more complete than Ferns, or Club-mosses,
although they are destitute of air vessels and breathing pores. Mosses are usually the first plants that
shew themselves on rocks, or walls, or barren places,
where no other vegetation can establish itself; provided the air is damp they will flourish there, and in
time, lay the foundation of a bed of vegetable mould,
in which the roots of grasses, and other stronger plants
may find support, till they, in their turn, have decayed and prepared the way for shrubs and trees.
This is the usual order observed by nature in converting the face of rocks into vegetable mould, and
thus you see Mosses have to perform the office of
pioneers to larger plants, an office for which one
would have thought their Lilliputian size would
hardly have qualified them.

Mosses are formed upon precisely the same plan as
flowering plants, as far as the arrangement of their
organs of vegetation. They have, in all cases, a
stem, or axis, however minute, round which the
leaves are disposed with the greatest symmetry—
(Plate XXIV. 1. *fig.* 2.); they have the parts that
answer to seeds, enclosed in a case, and this case is
elevated on a stalk, which arises from among the
leaves. But, beyond this, analogy ceases; in all
other points of structure, the Moss tribe is of a most
singular nature.

Mosses are said to be in fruit when the stems are
furnished with brown hollow cases, seated on a long
stalk (*fig.* 1. 2.). It is chiefly of this fruit, or theca,
and its modifications, that we make use in distin-

guishing the genera. Let it, therefore, engage our
attention the first. No species can be more common
than *Wall Tortula* (Tortula muralis), dark tufts of
which are found every where upon the north side of
walls, growing out of the mortar. The theca of this
plant wears a little cap, not very unlike that of the
Norman peasant women (*fig.* 2. *a.*), with its high
peak and long lappets ; this part is called a CALYP-
TRA ; when young it was rolled round the theca, so
as completely to cover it over like an extinguisher,
but when the stalk of the theca lengthened, the calyptra was torn away from its support, and carried up
upon the tip. After a certain time, the calyptra drops
off; and, at that time, the theca is in the best state
for examination. You will find it terminated by a
conical lid (*fig.* 3. *a.*), or OPERCULUM, which is thrown
off when the spores, or reproductive parts, are fit to
be dispersed (*fig.* 4. *b.*). When the lid has been thus
spontaneously thrown off, a new and peculiar set of
parts comes in view; you will find that the lid covered
a kind of tuft of twisted hairs, which at first look as
if they stopped up the mouth of the theca. But,
if you cut a theca perpendicularly from the bottom to
the top, you will learn from the sectional view that
you will then obtain of the parts, that, in reality, the
hairs arise from within the rim of the theca, in a
single row (*fig.* 5.). These hairs are named in Botany, the TEETH of the fringe, or PERISTOME ; the latter term designating the ring of hairs. The nature of
the fringe varies in different genera ; sometimes it
consists of two rows of teeth, differing from each

other in size or number, or arrangement; some have
only 4 teeth, others 8, or 16, or 32, or 64; in all
cases, their number is some multiple of four; a curious circumstance which shews the great simplicity of
design that is observed in the construction of these
minute objects. The fringe is not, however, always
present; there is a small section of the Moss tribe,
all the genera of which are destitute of that singular
organ. What office it may have to perform, we can
only guess; it seems to be connected with the dispersion of the spores, and often acts in the most
beautiful hygrometrical manner. If you take the
theca of this Tortula, for example, when dry, and
put it in a damp place, or in water, its teeth will un-
coil, and disentangle themselves with a graceful
and steady motion, which is beautiful to look upon.

It is in the inside of the theca that the spores are
confined. They lie there in a thin bag, which is
open at the upper end, and which surrounds a central column, called the COLUMELLA. They are exceedingly minute, and not unlike the spores of Ferns.

A superficial observer would remark no further organization than this; but the accurate investigations
of Botanists, have led to the discovery, that there
is a more minute and concealed system of organs,
which, in many cases, precedes the appearance of the
theca. It has been thought, that these organs represent pistils and anthers of an imperfect kind; but
you will see that, if they are to be so understood,
they are so much more imperfect, and so differently
constructed from the parts that bear those names in

flowering plants, as to render it extremely doubtful whether they can really be considered of the same nature.

At the end of the shoots of some Mosses such as *Hairmoss* (Polytrichum), the leaves spread into a starry form, and become coloured with brown. Among them lie a number of cylindrical whitish-green bodies (*fig.* 7.) which are transparent at the point, and filled with a cloudy granular matter, which it is said that they discharge with some degree of violence. These are considered to be anthers. But they appear to exist in some Mosses only, and not to be universally present, as they would be if they were really necessary to produce fertilization.

The second kind of apparatus is universally provided, and is in reality the forerunner of the theca. In the bosom of other leaves you may find, a short time after Mosses have begun to grow, a cluster of little greenish bodies (*fig.* 6.), which are thickest at their lower end, then taper upwards into a slender pipe, and at last expand into a sort of shallow cup. After a certain time the pipe and the cup, which, by some, are considered style and stigma, shrivel up, and the lower part, or ovary, swells, acquires a stalk, and finally changes into a theca.

The study of the distinctions of Mosses requires great care and attention, and much skill in the use of the microscope. It has sometimes occupied the undivided attention of Botanists, and cannot be attended to without much leisure and patience. The best work we have upon the subject is the Muscologia

Britannica of Drs. Hooker and Taylor, a valuable book, in which are accurate figures of nearly all the species found in this country. To it I refer you, if you wish to prosecute this branch of Botany.

Of about the same rank in the scale of organization are the plants called *Jungermannias*, which look very much like Mosses (Plate XXIV. 2. *fig.* 1.), and which like them have little roundish bodies called anthers (*fig.* 4 & 5.) and a theca *(fig.* 2. *c.*) elevated on a stalk. They are distinguishable; firstly, by their theca bursting into valves, and secondly, by their spores being mixed with tubes formed of curiously twisted threads *(fig.* 9.), and called ELATERS. They grow in tufts and patches, in damp and shady places all over Great Britain, occupying the bark of trees, and the surface of rocks and stones, or creeping among the herbage on the banks of rivers, on heaths, marshes, and in shady woods, and even inhabiting gloomy caverns where scarcely any other vegetable can exist. A noble illustration of these tiny plants was published some years ago by Professor Hooker; it forms the most complete local monograph of any genus ever published, and is indispensable to all those who would occupy themselves with an inquiry into the habits and differences of the tribe.

281

EXPLANATION OF PLATE XXIV.

I. The Moss Tribe.—1. A patch of *Wall Tortula* (Tortula mu-ralis) of the natural size.—2. A single plant magnified ; a calyptra.—3. A part of a theca with the lid a.—4. A theca from which the lid b has been removed ; a the twisted fringe.—5. A portion of the theca viewed from within ; shewing the teeth of the fringe.—6. A cluster of the young thecæ of Encalypta vulgaris, intermixed with succulent fibres.—7. A cluster of the bodies supposed by some to be anthers, in the same plant. —The two last figures are after Hedwig.

II. The Jungermannia Tribe.—1. A plant of the *Grove Junger-mannia* (J. nemorosa), natural size.—2. A portion of the same magni-fied ; a leaves ; b the sheath of the theca, sometimes called the calyx ; c the theca emitting its spores.—3. A leaf, bearing little warts, a a, which break up into reproductive particles b; and are called *gemmæ*.—4, 5, 6. One of the bodies supposed to be anthers, in different states, with the succulent filaments that are intermixed with them. —7. The calyptra of the theca, a little broken at its side.—8. Imperfect young theca.—9. Two of the elastic elaters, with the spores that are mixed among them. —These figures are copied from Dr. Hooker's British Jungermanniæ, Tab. 21.

LETTER XXV.

THE LICHEN TRIBE—THE MUSHROOM TRIBE— THE SEAWEED TRIBE.

(Plate XXV.)

AT last we have nearly reached the limits of the vegetable world, and find ourselves upon the confines of that mysterious region where animal and vegetable natures become so blended, that the philosopher is obliged to admit the weakness of his systematical distinctions, and to abandon the attempt at drawing a line between vitality and volition. In explaining to you the structure of the three lowest tribes in the vegetable kingdom, I do not propose to fatigue your attention by the minute, and specific details in which they so peculiarly abound; but I shall content myself with sketching out the great features by which they are usually distinguished.

When we quit Mosses and Jungermannias, with the other plants of a similar nature, we find ourselves among beings in which all trace of stems and leaves have disappeared, and which consist of nothing but thin horizontal expansions of vegetable matter in which a few harder, and differently formed kernels or shields are imbedded. In some of these the colour

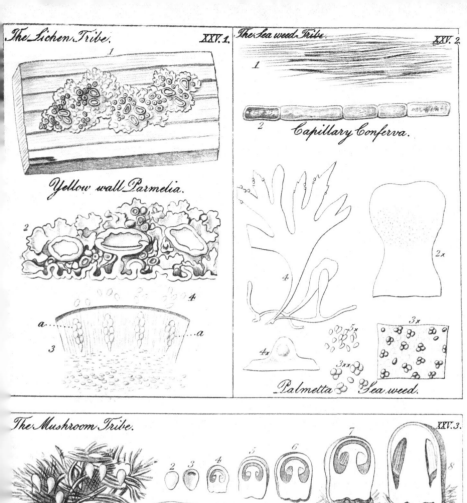

The Lichen Tribe. XXV. 1.

Yellow wall Parmelia.

The Sea weed Tribe. XXV. 2.

Capillary Conferva.

Palmetta Sea weed.

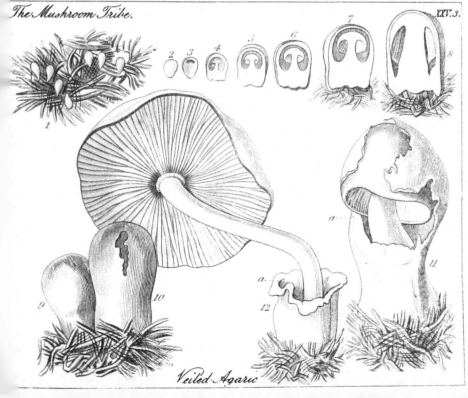

The Mushroom Tribe. XXV. 3.

Veiled Agaric

is yellow, brown or green, the texture of the expansion leafy, and the margin cut up into many lobes; these are the most nearly related to leafy and more perfect plants (Plate XXV. 1. *fig*. 1. & 2.); in others the expansion is merely a thin crust, which readily crumbles in pieces, the species having scarcely vital energy enough to keep the cells of which they are composed in a state of cohesion. Such plants as these are called *Lichens*. They are found chiefly in the temperate or colder regions of the earth. Some of them (Gyrophoras) crawl upon the surface of the earth, spreading their dingy, cold, and damp bodies over whole plains in the desolate regions of the north; others (Usneas, Ramalinas, &c.) spring up on the branches of trees, and hang down from them like grey and netted beards, giving the unfortunate plants of which they take possession, a hoary wintry aspect even in the summer; some (Parmelias, Lecideas, &c.) overrun old walls, stones, and rocks, to which they communicate those mild and agreeable tints, which render ancient ruins so peculiarly pleasing to the eye; and finally a fourth description of Lichens (Opegraphas) establish themselves upon the bark of living trees, occasionally burying themselves beneath the skin, through which their shields alone peep forth in the strange form of the letters of some eastern tongue.

Plants of this tribe have no parts in the smallest degree resembling flowers; they have no certain mode of multiplying themselves except by the dispersion of little spores, which are nothing but exceedingly minute cells that are lodged in the centre of the shields. These are very difficult to find; you may, however, make them out if you observe the following directions. Take the full grown shield of any Lichen; that of the *Yellow wall Parmelia* (P. parietina) is a good one for the purpose (Plate XXV. 1. *fig.* 2.); with a sharp razor divide it perpendicularly; then shave off the thinnest possible slice of one of the faces, and drop it into water; place it on the glass stage of a microscope, and illuminate it from below. You will then be able to perceive that the kernel consists of a crowd of minute compact fibres, planted perpendicularly upon a bed of cellular substance (*fig.* 3.); and that in the midst of the fibres there is a great number of little oblong bags (*fig.* 3. *a.*) filled full of transparent cells (*fig.* 4.); the bags are thecæ, the cells are spores; and it is to the latter that the Lichen has to trust for its perpetuation.

The study of Lichens is probably the most difficult of any part of Botany; the species are scarcely to be distinguished, the limits of the genera are uncertain, and the characters by which they are separated are obscure. If, however, you are curious to make yourself acquainted with these, the best book I know of to recommend to you, is the fifth volume of the English Flora, by Professor Hooker.

Notwithstanding their minuteness and uninviting appearance, several of them are of considerable importance to man and animals. The Arctic *Gyrophoras*, called by the Canadians *Tripe de Roche*, were the only food that the daring travellers Franklin,

Richardson, and Back, were for a long time able to procure in the horrible countries they so fearlessly visited in the cause of science ; *Reindeer Moss* (Cladonia rangiferina) is the winter food of the Reindeer of the Laplanders ; *Iceland Moss* (Cetraria islandica) furnishes a nutritive food to the invalid ; and finally the production of *Orchil*, by Roccella tinctoria is an indication of the value of some species to the manufacturer as dyes.

Very closely related to Lichens, and standing almost parallel with them in the scale of organization, are *Fungi*, or the *Mushroom tribe* (Plate XXV. 3.), plants with which we are best acquainted from our knowledge of the common eatable Mushroom, but which have almost an endless diversity of form and organization. In this, however, they are all agreed, that while the thecæ of Lichens are placed in shields or receptacles, which are exposed to the air, those of Fungi are in all cases concealed by a covering of some kind.   To give you any thing like an accurate notion of the many different appearances of Fungi, would be wholly impracticable in a work of this kind ; Dr. Greville's Scottish Cryptogamic Flora should be your guide, if you would dip deeply into the mysteries of their organization.   They vary from simple cells that hardly adhere, to chains of cells which resemble a necklace, thence to hollow balls, infinitely minute, that are generated in the living substance of leaves and stems, which they afflict under the names of mildew or blight, these again are developed in subterranean masses of cellular sub-

stance, such as the Truffle, and finally arrive at their
most perfect state in the Agarics, or Mushrooms, that
we eat, and in the Boleti, which grow like huge
fleshy excrescences on the trunks of trees, or project
from their trunks in long and ugly lobes, which have
in one case been compared to the claws of some gigantic demon.

Instead of occupying you about the particulars of all
these matters, I shall content myself with explaining
to you the nature of the developement of one of the
most completely formed Fungi, *the veiled Agaric* (Agaricus volvaceus). In the beginning this plant is nothing but a thin layer of cobweb-like matter, spreading among old tan ; by degrees, on the surface of the
cobwebs appear little protuberances of a whitish
colour (Plate XXV. 3. *fig.* 1.) ; they gradually
lengthen, and acquire a sort of stalk, and up to a
particular period consist of only a fleshy mass of
fibres and minute cells ; if they are cut through
at that time in a perpendicular direction (*fig.* 2.),
they present one uniform face. But in a short time
a minute cavity is formed in the fungus at the thicker
end (*fig.* 3.), within which a sort of cap is gradually
elevated upon a stalk (*fig.* 4.) ; the cap and stalk
keep progressively enlarging (*figs.* 5, 6, 7, 8.), and
stretching the skin within which they are enclosed,
till at last the skin cracks (*fig.* 10.) ; the cap and its
stalk rapidly enlarge, and tear a way through the
skin (*fig.* 11.), and at last burst forth into light, a
perfect mushroom (*fig.* 12.), with numerous cinna-
mon brown *gills* or LAMELLÆ radiating from the stalk

underneath the cap, and concealing the thecæ in
which the spores are laid up. When the Mushroom
has gained its full size, its stalk is surrounded at the
base by a thick fleshy sheath, called the VOLVA, or
wrapper ; from what you have seen of its gradual
progress, you will have observed that the wrapper is
nothing but the remains of the skin within which the
fungus was formed. I send you a copy of a beautiful
illustration of these phenomena by Professor Th. Nees
v. Esenbeck, of Bonn, by whom they were observed.

So simple is the growth of this and such plants,
and so remarkable are the circumstances under which
they are formed, as to have given rise to the belief
that they are not propagated by the agency of spores,
which are always sure to reproduce the plant from
which they originate, but are dependent for their
peculiar appearances upon the different circumstances
under which they are developed. Professor Nees
v. Esenbeck has proved that whatever error there
may be in such opinions generally, they are at least
well founded in the case of the Veiled Mushroom ;
for he ascertained, by careful observation, that the
same cobwebby matter which gave birth in hothouses
to the Mushroom, in long and bright days, when
there is plenty of light, produced nothing but a plant
called Sclerotium Mycetospora, in the autumn, winter,
and spring, when the hothouses abounded with heat
and moisture, but when the days were short, the sky
cloudy, and light deficient.

The last and lowest of all the tribes of plants are
the *Sea-weeds* and their allies ; these productions,

which inhabit water exclusively, and appear at one end of the scale of their developement in the form of enormous Fuci, many fathoms in length, but at the other as merely simple bladders sticking together in rows, are those to which I referred at the beginning of this letter, as forming the link between the Animal and Vegetable worlds. Like Lichens and Fungi, they have reproductive organs of the most simple construction; in those species which have the most complex organization, the spores are stored up in peculiar receptacles, as in the larger and more perfect sea-weeds; but in others they are distributed vaguely through the whole substance of the plant, and start into life when liberated from their nests by the destruction of the individual that generated them. In the *Lavers*, whether of fresh or salt water, they lie clustered in threes or fours, in the substance of a green membrane; in the true Confervæ they are nothing but granular matter, locked up in little transparent tubes (Plate XXV. 2.). It is of a vegetation of this latter kind that consist the green slimy patches you see floating in water, or adhering to stones and rocks from which water has receded In Dr. Greville's Algæ Britannicæ, and in the fifth volume of the English Flora, you will find a full account of the genera and species of these singular productions.

What is most remarkable in them is their approach to the nature of animals, an approach which is not only indicated by the apparently spontaneous motions of the kinds called Oscillatoria, but in a much more unequivocal manner by other kinds, if we can

believe the concurrent testimony of several French
and German Botanists. No one has investigated the
subject with more unwearied assiduity than Mons.
Gaillon, from whose "Observations sur les limites
qui séparent le regne Végétal du regne Animal," I
shall translate some details that I think cannot fail to
amuse and surprise you.

On the rocks that are found at low-water mark on
the coasts of Normandy and Picardy, there grows a
production called by Botanists *Conferva comoides;* it
consists of fine brownish-yellow threads, collected in
the form of a hair-pencil, half an inch or an inch in
length, and at low water spreads over the surface of
the little round calcareous stones, to which it gives
something of the appearance of the head of a new-
born child. These threads are loosely branched, and
are finer than the most delicate hair; the plant owes
its apparent solidity to the clustering and entangle-
ment of many such threads. Viewed under a microscope that magnifies 300 diameters, the threads
seem to be rounded, slightly compressed, and
about as large as fine packthread. They are of a
mucous nature, and contain immersed within their
substance a number of small yellowish bodies, which
look at first like dots, afterwards become oval, and
end in acquiring something the shape of a radish,
having the ends transparent, and the centre marked
by a patch of yellowish matter. If they are at that
time separated from the mucous matter in which they
are pressed and packed like herrings in a barrel, you
may see them moving, expanding, contracting, ad-

vancing gravely and slowly, retreating in like manner, altering their direction, and finally possessing a
spontaneous, incessant, measured, voluntary motion.
These little creatures, which at most are not more
than the 1000th of an inch long, and at the smallest
hardly exceed the 5000th, when once they are separated from the thread that contains them, fall down
in countless multitudes, in the form of a chocolate-
brown deposit, on the neighbouring rocks.   Once
there, they distend, and emit a globule of coloured particles, which are evidently their fry. Each
particle gains motion and volume, and the little globular mass, lengthening and branching, reproduces,
by the developement of the germs that are collected
together, the long green pencilled appearance, which
has led Botanists to consider this being as a plant.

In another production, *green ditch-Laver* (Ulva bul-
lata, or minima, or Tetraspora lubrica), still more
astonishing circumstances have been observed by M.
Gaillon and others. This plant appears, to the naked
eye, a thin green membrane, within which the microscope reveals a number of green granules, arranged in fours.   Let this membrane be kept in quiet
water, and at a high atmospheric temperature, and the
granules may be seen, under a powerful microscope,
to present at their surface certain convexities and depressions, which are the effect of the repeated con-
traction and distension of these granules.   If they are
carefully watched for several days, the granules will
be seen to be reciprocally displaced ; after a certain
time they separate from the membrane, and may

then be perceived to have a rapid and regular movement, as if in chase of each other; cool with a drop of water, that in which the granules are floating, and their motions will become slower, they will attach themselves by some part of their circumference, and will acquire a swinging motion from right to left and from left to right. In this sort of imperfect reeling and twirling, one sees the granules approach in pairs, just touch each other, retreat, approach again, and glide away to the right or left, staggering, as it were, and trying to preserve their balance; at last, instead of pairs, fours combine to execute the movements of the dance. Imagine the field of the microscope covered, shortly after, with a hundred of these animated globules, whose diameter is not, in reality, more than the 4000th of an inch, chasing each other, retreating, and intermingling, as if executing the mazes of a fantastic reel, and you have one of the most curious spectacles that the microscope can exhibit. When great numbers of the granules are collected, the motion ceases; they then collect in fours, and form a new membrane, and in this state they are considered by Botanists as a kind of vegetable.

Such are, in part, the wonders revealed by the microscope in these ambiguous productions; many others of equal interest might be named, but what has been said will suffice to shew you how marvellous a store of curious facts remains to be collected by those whose time and disposition are favourable to such inquiries. To these may be applied

the lines of the French poet Delile, even better than
to the zoophytes to which allusion was more particularly made :—

> Voyez vous se mouvoir ces vivans arbrisseaux
> Dont l'étrange famille habite dans les eaux,
> Et qui, de deux états nuance merveilleuse
> Confondent du savoir l'ignorance orgueilleuse.
> De l'humile séjour ces douteux habitans
> A l'œil inattentif échappèrent long-temps ;
> Ils vivaient inconnus, et sujets de deux mondes,
> En se multipliant voyageaient sur les ondes.

EXPLANATION OF PLATE XXV.

I. The Lichen Tribe.— 1. A specimen of the *Yellow Wall Parmelia* (P. parietina), growing on a piece of pale.—2. A section of the
same magnified.— 3. A perpendicular section of a shield ; *a a*, thecæ,
with the spores in them.—4. The spores.

II. The Sea-weed Tribe.—1. A mass of *Capillary Conferva* (C.
capillaris).—2. A thread highly magnified.—1*. *Palmetta Sea-weed*
(Rhodomenia palmetta), natural size, with its masses of reproductive
matter.—2*. A portion of the plant with reproductive particles collected
in clusters of 3 or 4.—3*. A portion of the last highly magnified.—
3**. Clusters separate.—4*. A bit of the plant with one of the masses
of reproductive matter.—5*. Spores.—The latter after Greville.

III. The Mushroom Tribe.—1. A patch of tan, on which young
plants of the *Veiled Agaric* (Agaricus volvaceus) are growing.—2—8.
Progressive developement of a young plant.—9. A plant just ready to
burst its wrapper.—10. The same bursting its wrapper.—11. A plant
emerging from the wrapper, *a.*—12. A perfect plant ; *a* the wrapper,
or volva.—(After Th. Nees von Esenbeck.)

I HAVE now conducted you through the whole kingdom of Vegetation ; and, I trust, that both yourself
and your children will have derived both amusement
and instruction from the journey. Of this, at least,
I feel assured, that the supposed difficulties of the
subject have vanished, and that the determination of
the more conspicuous and important of the great
tribes into which plants are naturally divided, is
placed completely within your reach. You must
have seen that they possess features too strongly
marked to be mistaken ; and that their characters,
divested of the technicalities of 'us Botanists, are
simple and precise.

I have, however, passed by many large Natural
Orders, with which I should have wished you to be acquainted, and of others I have only made casual mention. To this I have been led, firstly, by a desire to
simplify your earlier studies as far as might be prac-
ticable; and, secondly, by the obvious impossibility
of comprehending every thing in a few elementary
letters. It will be found, however, that the whole system turns upon an acquaintance with the great forms
of Vegetation I have actually illustrated, and that
nearly all the other Natural Orders have an obvious
resemblance to one or other of those you have seen,
which you may consider as so many resting-places,
from which you may proceed to the consideration of
the distinctions between them and others you have
hitherto heard of imperfectly, or not at all. If you wish
to pursue the study beyond the general principles of
the Natural System, you will, of course, provide

yourself with the books in which the mutual relations of all plants are explained ; if you have no such intention, you cannot experience inconvenience from the want of a more extended and universal knowledge of the subject.

For myself, it would have been a pleasure to me to have continued my explanations till the subject was quite exhausted ; but many causes concur to render it impossible for me to accompany you further in person. In the remainder of your progress in pursuit of Botanical knowledge, if, indeed, you purpose to continue it further, as I most sincerely hope you do, you must rest satisfied with my best wishes for a prosperous result to so laudable an undertaking.

INDEX.

THE NUMBERS REFER TO THE PAGES.

302

INDEX.

THE END.

NORMAN AND SKEEN, PRINTERS, MAIDEN LANE, COVENT GARDEN.

Printed in the United States
By Bookmasters